CFZ YEARBOOK 1999

Typeset by Jonathan Downes,
Cover and Layout by le chat d'orange for CFZ Communications
Using Microsoft Word 2000, Microsoft , Publisher 2000, Adobe Photoshop CS.

Photographs © 2008 CFZ except where noted

First published in Great Britain by CFZ Press

**CFZ Press
Myrtle Cottage
Woolsery
Bideford
North Devon
EX39 5QR**

2nd Edition © CFZ MMVIII

All rights reserved. Without limiting the rights under copyright reserved above, no part of this publication may be reproduced, stored in or introduced into a retrieval system, or transmitted, in any form of by any means (electronic, mechanical, photocopying, recording or otherwise), without the prior written permission of both the copyright owners and the publishers of this book.

ISBN: 978-1-905723-24-9

Contents

3. Contents
5. Introduction to the 2007 edition
9. Introduction to the original edition
11. *Surviving Neanderthals - an overview of the search for man's closest relatives* by Jonathan Downes and Richard Freeman
49. *Pelorus Jack* by James Cowan (a reprint of the 1912 booklet)
73. *The Beast of Bluebell Hill* by Neil Arnold
89. *Waitoreke - The enigma from New Zealand* by Craig Heinselman
105. *In the Shadow of Wolf's Castle* by Roy Kerridge
117. *Did some Mesozoic reptiles evolve from flying bir*ds by Allan E Munro
121. *A Ghana Folk Tale* by Louis Baba
133. *Is there anybody out There?* by Graham Inglis
143. 1998 - A year in the life of the Centre for Fortean Zoology

INTRODUCTION
to the new edition

I have learned quite a lot while carrying out this long overdue repackaging of the CFZ yearbooks. Most of what I have discovered has been quite a positive experience for me - many people (including, I blush to admit, me) have considered that the CFZ only really got going in about 2002, when we carried out our first big high profile expedition to Martin Mere, and solved the mystery of the `monster` that dwelt therein.

When I wrote my autobiography a few years ago I glossed over the years between 1996 and 2002, because they were particularly horrible ones from my point of view, and they were times that I didn't really feel like revisiting. I went through a horrible divorce in 1996, and for the next six years I did my best to drink myself to death. I also had a high profile drug problem, and both physically and emotionally I was a mess.

I always felt a certain amount of empathy with guitarist Joe Walsh, who once claimed that he had no memories at all of certain periods of his life because he had been so stoned for most of the time, that they had just rushed by in a blur. I wasn't quite that bad, but a mixture of substance abuse, mental health issues, and personal problems meant that for some years my life was so horrible that I have done my best to forget about it.

Re-reading the issues of *Animals & Men* which I recently remastered for the third of our volumes of collected editions, and now this yearbook, I can now look back upon those days with a certain degree of greater enthusiasm, as I realise that even then, the CFZ was doing a remarkable job. All three of us were on the dole, and had practically no income, and it remains somewhat of a mystery how we actually managed to keep up such a hectic publication schedule, but we managed it.

I have always disliked a mindset that seems to be prevalent amongst many authors of practically rewriting their books each time they are reissued. I have bought new editions of many old favourites to find that they have been edited so severely that they are practically unrecognisable from their original editions. As CFZ Press have worked their way through the programme of reissuing our back catalogues as perfect bound paperbacks, rather than in their original spiral bound and photocopied form, I have always made sure that the team has stuck to this ethos, and resisted the temptation to rewrite history.

However, with this volume we have had to make an exception.

This was the fourth of our yearbooks, and was the first one to be put together on a PC. When a small publishing company asked to print a new edition of my book *The Owlman and Others* they paid me an advance of a couple of grand. It says something about how the price (and processing power) of computers has plummeted since then, that when - in November 1997 - I bought my first PC it cost me nearly a thousand pounds, and had a whopping great gigabyte of hard drive.

However it took me a long time to work out how to use it, and whereas *Animals & Men* was produced on a PC from issue 16, the 1998 yearbook was produced as its predecessors had been, with the aid or letraset, glue, and my trusty Amiga 500+. With the 1999 yearbook we finally bit the bullet and produced our first book on it! With hindsight it was a complete shambles, and it is with a great deal of embarrassment that I discovered quite how many typographical errors, and plain cock ups there were inside its pages.

The first three Yearbooks (1996-8) have been reproduced in their new editions in purely facsimile form, but as we still have this volume in electronic format, we decided that it really would benefit from being re-edited. (We shall be doing the same for the later volumes in the series; 2001-2, 2003, and 2004).

The later volumes in the series are nowhere near as scandalously badly edited as this one was, but in our own defence I would like to explain why these appalling mistakes got through.

Although our many vociferous detractors would - I am certain - like to say that it was because Graham, Richard, and I were too drunk or stoned at the time to do a proper job, this just simply isn't the case. The truth is that we got carried away with our new technology and leapt in with both feet before we really knew what we were doing. We didn't realise (for example) the difference between a word processing programme and a desktop publishing one, and we had no idea that spell checkers existed. We also - or so, by laboriously going through our work logs for the time, it appears - managed to delete the spell-checked proof version, and do the design with the unedited one. We even managed to leave out one of the articles from the Contents page.

It was also a case of too many cooks spoiling the broth. We all wanted to have a go at using the new technology. As well as the three of us, there was a motley band of hippies and Exeter low life who at the time seemed to congregate at the CFZ, and several of them, got involved as well. When one takes a long, hard, look at the conditions under which the 1999 yearbook was produced, it is a miracle that the eventual volume was as good as it was.

I had never realised quite how bad the production values of the original volume were, or else it would have been remastered years ago.

But there is no use in crying over spilt milk. The production of the original edition of the 1999 edition was an important part of the CFZ learning curve, and if we hadn't taken our few faltering original steps with DTP programmes we would not be where we are today - unarguably the biggest cryptozoological publishing house in the world, in only nine short years.

We have taken the opportunity to fix as many spelling erors, and typos as we could find, and have enhanced the original illustrations wherever possible, as well as adding a number of new ones. We have also added a number of footnotes, because - although we still adhere to the belief that Stalinist rewrites and rejigglings of one's back catalogue is a particularly un-aesthetic thing to do, (I still remember with horror buying a `remastered` copy of Frank Zappa's *Only in it for the money* (1967) only to find that the Zappatista had - in his wisdom - decided that it would be a pretty groovy thing to do to rerecord all the rhythm sections with then-current technology, and his late 1980s touring band, with disastrous results), quite a few important things had changed in the nine years between the original edition and the new one. People (Quentin Rose, and Bernard Heuvelmans in particular) had died, and statements like "The CFZ hope one day to carry out an expedition to Mongolia" were nearly half a decade out of date. Such changes, however, have been limited to footnotes, in order to adhere as closely as we could to our original aesthetic sensibilities.

We have not changed the original text in other ways, even when it seemed dated. For example, when we wrote our original paper on the viability of hypothetical surviving Neanderthals, the collapse of the Soviet Union was well under a decade previous, and had been one of the defining moments of our adult lives. Re-reading some passages now - nearly 20 years after Gorbachev presided over the dismantling of the USSR, one would probably not these days describe Kazakhstan (for example) as a "former Soviet Republic", as there is now at least one full generations of adults for whom there has never been any other state of affairs.

However this book is of its time, and I would not wish to change that.

Here I would like to say a few words about the publishing schedule for the CFZ Yearbooks, although, if everything goes according to plan, in a few short months these next few paragraphs will be of academic interest only.

I first came up with the idea of the CFZ Yearbook series in the autumn of 1995 when I will freely admit that I wanted to emulate the success of the then popular *Fortean Studies* series which for some years was published by those jolly nice folks at *Fortean Times*. The first volume came out before Christmas 1995, and therefore had a shelf life for the next thirteen months. This was a feat that we were not to emulate again until 2007 when the 2008 yearbook came out just before Christmas once again.

We published the yearbooks annually until 1999, but there was no yearbook in 2000. This (again I will freely admit) was because of my deteriorating health, both physical and mental. However, in 2001 we were back, with the 2000/1 yearbook, and as both my condition and that of the CFZ improved, we published new editions in 2002, 2003 and 2004.

Then Richard's father died, and mine became seriously ill with the condition that in 2006 was to kill him, and we both had to take time out for family reasons. There were no yearbooks in 2005 and 2006, but we returned in 2007, and because - on this occasion - there was no need to be embarrassed about the reason for our absence, we named it the 2007 Yearbook.

It would have been much easier if we had carried out the reissues in linear order, but that was just simply not feasible. We spent much of 2006, and the first months of 2007 reissuing our back catalogue, but we felt that we wanted to move on, and for most of 2007 we concentrated on issuing new cryptozoological titles.

However, as we sold out of the original editions, we republished them in perfect bound form. The first was the 2004 Yearbook, which came out in the autumn of 2007. The second was the 1996 Yearbook which came out a few weeks ago, and this volume is the third.

The 2003 volume will be next, and then the 1997 and 1998 ones. These two will be in pure facsimile form, as they masters do not exist electronically. The final two volumes will be 2000/1 and 2002 later in the year.

Historically, by the time we got to 2003 we knew what we were doing a lot more than we did back in 1998/9 and so although the 2003 volume (and the 2000/1 and 2002 ones, when they get done in a month or two), will be re-edited to remove embarrassing typos, and footnotes will be added where appropriate, I strongly suspect that they will not take as long as this volume.

But I don't want to sound like I am being overly negative either about the original edition, or about the work that has been involved with producing this new version. I am very proud of what we have achieved with the CFZ, and the 1999 Yearbook was an important step along the way.

Onwards and Upwards,

Jon Downes
Director, CFZ
(Woolfardisworthy),
North Devon
February 2008

INTRODUCTION
to the original edition

Dear Friends,

It seems incredible that this is our fourth (yes, count them!) Yearbook. Once again despite the fact that we are later than we thought that we would be, we have managed to produce a book of which, I feel, we can be justly proud.

This year there are fewer articles than in previous issues, but they are longer than they have been in other years - indeed several of the articles are of the length that some publishers feel like putting out as complete books! This only goes to show that The Centre for Fortean Zoology may be slow, but we get there in the end, and that on the whole, when we do so it was worth the wait This year the subject range is more diverse than it has been previously, but it is nice to welcome new writers, (like Craig Heinselman and Graham Inglis) to join the established Yearbook contributors like Richard Freeman and Roy Kerridge.

We make no apologies for having reprinted the entire text of a 1912 booklet about *Pelorus Jack* the semi-legendary New Zealand dolphin. The booklet was only available in New Zealand, it has been out of print for yonks and anyway, we felt like it. I think that when you read this extraordinary account, especially in conjunction with the chapter on `Beaky` the dolphin of Falmouth bay in my book *The Owlman and Others* (Domra 1998) it gives one quite an interesting insight into the interaction between humans and Cetacea.

In Issue 11 of *Animals & Men* CFZ deputy director Graham Inglis promised a new series on Exobiology. It didn't happen, but in this volume is an extremely scholarly treatise on the subject which lends weight to the supposition that the truth is indeed out there.

I hope that you all enjoy this, the last yearbook of the 20th Century and that you will continue to support us for another year....

Jon Downes (Director)
The Centre for Fortean Zoology

Surviving Neanderthals?
An overview of the search for our closest relatives

by
Jonathan Downes and Richard Freeman

"Man is the animal who laughs"
Michael Valentine Smith

According to the Declaration on Race and Racial Prejudice adopted and proclaimed by the General Conference of the United Nations Educational, Scientific and Cultural Organisation at its twentieth session, on 27 November 1978: *"All human beings belong to a single species and are descended from a common stock. They are born equal in dignity and rights and all form an integral part of humanity"*. Whereas the ideology behind this statement is unquestionably sound, and it is not the intention of the present authors to question UNESCO or its stand against racial discrimination, we believe that the above statement is completely wrong. *Homo s. sapiens* is NOT the only species of man to live on Planet Earth.

In August of 1856 - three years before the publication of Darwin's *On the Origin of Species* sent figurative shockwaves across the scientific establishment - a party of German labourers in search of lime blasted out the entrance to a small cave (the Feldhofer grotto) that lay high on the sheer wall of the Neander Valley (in German Neander T(h)al, near Düsseldorf, Germany, through which the Düssel river meanders to join the Rhine.

In the debris from the explosion, they found part of a skull, which was unlike anything that had been seen before. It was long and low, with a pair of large ridges arching over the now-vanished eye sockets. Nearby they excavated some bones from the body of the same heavily fossilised and very robustly built individual. The workers did not think anything much of these finds, assuming them to be the bones of a cave bear; but by great good fortune they set at least some of them aside for eventual examination by the local schoolteacher and amateur natural historian Johann Fuhlrott. Fuhlrott, to his eternal credit, had the insight to recognise them for what they were: the remains of a previously unknown type of human. Fuhlrott took the finds to Hermann Schaaffhausen, professor of anatomy at the University of Bonn, and after a preliminary announcement by Schaaffhausen, the pair presented the Neanderthaler ('Neanderthal Man') to the world at a meeting of the local natural history society in June 1857.

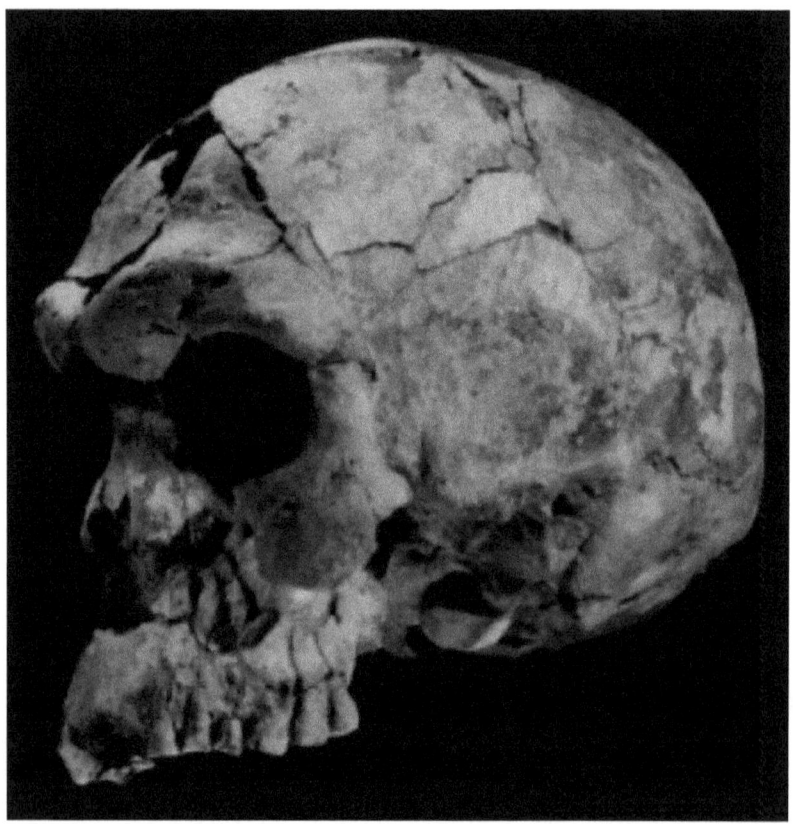

This was the first evidence of a distinct species or subspecies of human, *Homo (sapiens) neandert(h)alensis.*. Sites at which (by now abundant) Neandert(h)al fossils have been found are distributed in Europe and western Asia from the Atlantic in the west to Uzbekistan in the east, and from Wales in the north, to Gibraltar and the Levant in the south. This is established fact, but according to scientific orthodoxy our primitive cousins died out - probably due to the predations of our own, more sophisticated species during the later part of the Pleistocene epoch, more familiarly known as the Ice Age, some 200,000 to 30,000 years ago.

Neanderthal Man has been much maligned by both scientists and lay-people, and it has only been within recent years that the truth has emerged about them. The commonly held view that Neanderthals were brutish and violent savages (the very word Neanderthal has become synonymous in the English language with football hooliganism and thuggery) has been shown up to be a complete fallacy. Between 1953 and 1957 the Columbia University archaeologist Ralph Solecki excavated the cave of Shanidar, in northern Iraq, recovering the remains of nine adult and juvenile Neanderthals. One of the skeletons was that of an adult male who had suffered, perhaps since birth, from a disease that withered his right arm. Solecki pointed out that this disadvantaged individual could not have survived to a relatively advanced age without the active and long-term support of his social group. Suddenly the Neanderthals were perceived as being caring and humane, as well as spiritually aware. This new Neanderthal persona was made yet more compelling by the discovery of fossil pollen that suggested the individual had been buried with spring flowers.

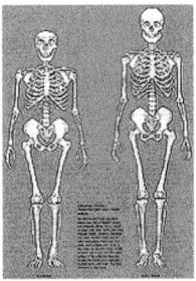

This illustration, done by Diane Salles, in Ian Tattersall's, The Last Neanderthal, *shows the differences that exist between modern humans and Neanderthal morphology! The Neanderthal is on the left, and the modern human is on the right. See how Neanderthals were slightly shorter than us, and their bone structure was more squat.*

The conception of *Homo s. Neanderthalis* as noble savages with a complex social structure, a religion and flower burials is a compelling one which has been exploited widely in novels such as *The Clan of the Cave Bear* by Jane Auel which describe a society with complex rituals who worshipped the Cave bear (*Ursus spelaeus*) - a huge creature that became extinct some 40,000 to 50,000 years ago.

Unfortunately it seems that the pendulum of perception had swung too far in the opposite direction because as compelling as the concept of Neanderthal Man as gentle proto-hippies living off the land is, recent evidence suggests that there is little reason to believe this interpretation of the available data either.

Johan M.G. van der Dennen, of the Centre of Peace and Conflict Studies, University of Groningen, the Netherlands, reviewed three recent books on the species and showed that each of the authors had extrapolated completely different conclusions from the same evidence, and reached the conclusion that nobody really knows the truth about our nearest relatives. The only thing that all the authorities he cited can agree on is that *H. s. neanderthalis* is now extinct, but even the manner of the extinction is a matter for debate. The jacket of Ian Tattershall's 1995 book *The Last Neanderthal: The Rise, Success, and Mysterious Extinction of Our*

Closest Human Relatives claims that:
"Weaving together the archaeological and fossil evidence with the lessons of evolutionary theory, Dr. Tattersall draws on our latest knowledge about how Neanderthals evolved and lived to solve the riddle of how they died. He presents convincing evidence to demonstrate quite conclusively that Neanderthals were killed off by invading Homo sapiens in the first known instance of human genocide."

Two years earlier, Stringer & Gamble published *In Search of the Neanderthals: Solving the Puzzle of Human Origins*, which presented a far more reasonable view of the extinction of Neanderthal Man:

"In an area as large as Europe, with its varied environments and over a timespan of perhaps 10 millennia, many different kinds of interactions could have occurred (and probably did occur), ranging from avoidance to tolerance to interbreeding, and from conflict and economic competition to friendship and an exchange of ideas... [very probably] there was minimal gene flow (interbreeding) between the two populations [because of] predominantly behavioural barriers that kept them distinct from one another...

If the Cro-Magnons became more skilled at coping with and exploiting the European environments than the Neanderthals, the Cro-Magnon populations and ranges would have increased. With only finite resources, the Neanderthals would have suffered from economic competition unless they withdrew to more marginal areas (such as, in this context, the southern Iberian and northern British peninsulae). If the Cro-Magnons occupied the more favourable and sheltered lowland valleys, the Neanderthals would have had to occupy higher or less-sheltered ground. In a normal summer this would have posed them few problems, but in more inclement weather their populations would have been put under severe stress.

They would have suffered from higher infant mortality rates and shorter lifespans. Repeated across various parts of Europe and over many centuries or even millennia, this attrition would probably have caused Neanderthal populations gradually to decline toward extinction.

In fact, using a computer-simulated model, archaeologist Ezra Zubrow has shown how rapidly that Neanderthals could have become extinct. Assuming interaction between stable populations of Neanderthals and Moderns, a Neanderthal mortality rate only 2 per cent higher than that of the Moderns could have resulted in Neanderthal extinction within about 1,000 years."

This model seems far more believable than Tattershall's blood thirsty vision of antediluvian ethnic cleansing and could well explain why the species vanished across what was quite an extensive range. However it does not explain why the species should have vanished completely, which in view of the burgeoning body of evidence that some Neanderthals have, indeed, survived to the present day, is probably a good thing.

We should start our search for surviving Neanderthals in the former Soviet Republic of Georgia, a small country on the eastern edge of the Black Sea. Even at the end of the 20th Century there are wild places which harbour a wide selection of large, and fierce animals. There are still bears and wolves, and the leopard still existed there within living memory. Ochamchir is a coastal region that has been in the news recently because of civil disturbances and ethnic unrest.

The Former Soviet Republic of Georgia

In the mid-nineteenth century, hunters from a village called T'khina that lies on the Mokva River and is roughly 50 miles from the city of Sokhumi, captured a 'wild woman' who had ape-like features, a massive bosom, thick arms, legs, and fingers, and was covered with reddish hair. This 'wild woman', named Zana by her captors, was so violent at first that she was housed in an enclosure of some type for a period of about three years. In this time she had very little human contact. Her food was thrown to her and she dug holes to sleep in. Eventually she grew calm around humans and was moved out of the enclosure, even given her freedom to roam about the countryside. After she became domesticated she would perform simple tasks, like grinding corn. She had an incredible endurance against cold, and couldn't stand to be in a heated room and for the whole of her life she refused to wear clothes.

She enjoyed gorging herself on grapes from the vine, and also had a weakness for wines, often drinking so heavily she would sleep for hours. As Colin Wilson points out in *The Encyclopedia of Unsolved Mysteries*, this is likely how she became the mother of many children to different fathers. She gave birth to at least six children, all sired by different men of the village who appeared to use her very much as a sexual plaything. The first two children died after Zana tried to wash them in the freezing river, a mistake that is understandable if she expected the children to have her own resistance to the cold. The villagers just started to take her children away from her and her "owner's" wife raised them as her own. Unlike their mother, the children developed the ability to communicate as well as any other villager.

At least four of her children survived. These were Dzhanda (male), born in 1878, Kodzhanar (female) 1880, born two years later, Gamasa Sabekia (female) 1882, and her youngest son Khwit Sabekia who was born in 1884. Zana died in the village about 1890; the youngest of her children died in 1954. Her story was researched by Professor Porchnev who interviewed many old people (one as old as a hundred and five) who remembered Zana, as well as two of her grandchildren. The grandchildren had dark skin and a Negroid look, and the grandson, named Shalikula, had jaws so powerful that he could lift a chair with a man sitting in it using only his mouth.

Although Zana's capture is far from being unique in the annals of cryptozoology - Dmitri Bayanov has collected together a number of accounts of captured wildmen - Zana is the only specimen on record who, because of her sexual proclivities, has left genetic material behind her. **

Sadly, for those people who have eagerly leapt on her `case`, as support for the theory of Neanderthal survival it looks as if she is no such thing.

When searching for Zana's true identity the first thing that we are able to do is prove what she was NOT. Although we have collected accounts of ape-like creatures from across the world for this book, no known species of animal can interbreed with human beings.

Zana, therefore HAS to be a hominid of some description. Professor Grover Krantz, a cryptozoologist specialising in hominoid survival has examined the skull of one of Zana's children. In *"Big Footprints"* (1992) he wrote:

"There is no direct evidence remaining of Zana herself, but a photograph of one of her sons shows nothing to indicate a less-than-human anatomy.

The remains of this man have been exhumed, and through the courtesy of Igor Bourtsev in Moscow I was able to examine the skull in great detail. It is a perfectly normal specimen of modern Homo sapiens, with slightly stronger jaws and more flare to the zygomatic arches (cheek bones) than is usual. anatomical evidence of this ancestry, and it is not there. Assuming

** A documentary aired in 2007 finally laid the Zana mystery to rest. It appears that Zana's skull was also exhumed some years ago, and that DNA testing upon it showed, not only that she *was* the biological mother of `Quit` her alleged son, but that Zana too was human.

Her DNA matched up within acceptable terms of reference with the DNA of modern human beings rather than with that of a Neanderthal.

However, this only goes to bolster up the theory that Richard and I proposed in this article, that Zana was an example of an extremely primitive human being - a Mesolithic era type hunter-gatherer, from a relict population of such people living in the remote parts of eastern Europe and the Causacus.

As we go to print with this present edition of the 1999 Yearbook, the CFZ are planning a trip to Kabardino-Balkaria for the mid summer of 2008, and - in liaison with Ukranian zoologist Grigoriy Panchenko - we hope that the expedition will be able to return to Britain with some conclusive evidence for the survival of primitive hominoids in the area. This is a very exciting project and we shall be bringing you news of the results as soon as we have them.

that this individual has been correctly identified, and I have no reason to doubt it, his anatomy would fit perfectly with one parent having been a late Mesolithic hunter"

There are plenty of scientific precedents for Krantz's interpretation of the story of Zana and her children. Each year "stone age" tribes are discovered living in the more remote parts of the world. During the preparation of this book Reuters News Agency reported that officials of the Indonesian Government had discovered two remote tribes on Irian Jaya:

"The report from the provincial capital Jayapura said that field officers of the social affairs office had recently discovered the two nomadic tribes living near the Mamberamo river area, about 2,400 miles east of Jakarta.

The office's head of social welfare, Onesimus Y Ramandey, said members of the Vahudate and the Aukedate tribes are tall, have dark skin and curly hair and speak using sign language. He said they roam the areas between Waropen Atas sub-district, Yapen Waropen district and the edge of the Mamberamo river in Jayapura district which borders with the Nabire, Puncak Jaya and Jayawijaya districts. Based on a preliminary survey, the Vahudate tribe has 20 families and the Aukedate tribe 33 families, he said.

Irian Jaya, with its high mountains, steep hills and deep valleys has hundreds of distinct tribal groups speaking up to 800 dialects with many still living far from the reach of government in a near Stone Age existence. New tribes are "discovered" almost every year".

Within a week another previously unknown tribe was discovered in The Amazon rainforest, and within the last five years similar have been discovered across Asia, South America and even Australia. The big question is, however, could a relict population of Mesolithic hunter-gatherers live undiscovered in Eastern Europe. The evidence suggests that the answer is a resounding YES. On March 29th 1992, "The Long Beach Telegram" carried a story about an expedition to the Causacus mountains between the Black and Caspian seas in search of the "Almasty" or Caucasian wildman:

"Dr. Marie-Jeanne Koffmann, a French-Russian surgeon, mountaineer and scholar, has been on the Almasty trail for more than two decades and has collected more than 500 accounts and a plaster-cast footprint of the 'forest man of the Caucasus.'

"She travelled on horseback through the remote mountains between the Black and Caspian seas, talking to villagers who had seen the mysterious beast. Although sceptical at first, she became convinced that the Almasty was another in an array of species that roamed the Caucasian wilds. Retiring in France on a tiny Soviet pension, she never dreamed that one day she'd have the money to mount a full-scale scientific search.

"But then, she had not counted on Sylvain Pallix. Pallix, a documentary filmmaker, was fascinated by two articles Koffmann wrote for Archologia *magazine. Tracking her down, he proposed finding sponsors for an expedition that he would film.*

"The respected French paleoanthropologist Yves Coppens gave the search his blessing. Pallix

raised half of the needed $1.8million. He's confident he'll find the rest. ``For three weeks, the telephone has been ringing off the hook,'' said Pallix, whose previous works have included a documentary on a Harley-Davidson meet in South Dakota and one on Calvados moonshiners. 'People are fascinated by the Almasty.'

A dozen people will leave Paris in June, to be joined by a dozen of Koffmann's scientific colleagues from Moscow. They will conduct their search in the Kabardin-Balkar region of Russia, just north of Georgia. The expedition hopes to find the beast, put it to sleep, take blood and skin samples and a plaster cast of the face and then let it awake in freedom - after putting a band on it so its wanderings can be followed.

"Appearing like a cross between an ape and a Neanderthal, the Almasty reputedly can run up to 37 mph. It is said to be omniverous and sometimes travels with companions and babies. The last sighting of the Almasty was by a zoologist friend of Koffmann who reported spending six minutes watching one on Aug. 25, 1991"

Whereas we reluctantly have to agree with Krantz's hypothesis that Zana, and other wildmen captured in Georgia and the surrounding countries are probably very primitive human beings rather than survivors of a relict population of Neanderthal Man, drawings by Jeanne-Marie Koffman which have been published in Dmitri Bayanov's *"In the Footsteps of the Russian Snowman"* (1996) suggest a Neanderthal identity for the elusive Almasty.

Kabardino-Balkarian Republic is a small Constituent of the Russian Federation which is situated in the Caucasus, that disputed area of far eastern Europe, which marks the eastern border with Asia, and which has always been within the Russian sphere of influence from Tsarist times to the present day. An autonomous province was established for the Kabardians in 1921 and a year later this was extended to include the Balkarians. In 1936, the status of the region was upgraded to that of an ASSR. This, in turn, declared itself a Union Republic in 1991.

In spite of the territorial proximity of the Caucasian peoples, their development took place in certain isolation due to the mountainous terrain. Kabardinians ethnically belong to the group of Northern Caucasian peoples. Their Kabardino-Cherkess tongue belongs to the Iberian-Caucasian languages of the Indo-European language family. The written language is based on Cyrillic. Balkarians are related to Turkic peoples and speak a Turkic type of language. Their written language is also based on Cyrillic.

Both Kabardinians and Balkarians are mostly Muslims (Sunnites). But it seems that there is a history of tension between the two peoples and indeed the Balkirs were deported between 1945-1954 for their collaboration with the Nazis during WW2 and during this time the republic was only inhabited by the Kabardin.

It is a wild and beautiful country about which surprisingly little is known. 70% of the land is inaccessible mountains - the largest of which is Mount Erbrus, the highest mountain in Europe - and, if true, Neanderthal survival has taken place anywhere in Europe it would be here.

Jeanne-Marie Koffman explored this area extensively, and during her travels met, Talib Kumyshev, 67, a Kahardian, who was a highly respected elder man of the village of Kamen-nomos. His testimony is highly revealing:

...It was probably in 1930, or 1931, or 1932, in June or at the end of May, when our cattle left for the alpine pastures of Elbrus. I was chief of the group. We had left to inspect the herds with the veterinarian.

Well, rain had surprised one of my shepherds, Shaghir Zagureyev, very high up on the slopes, and he had gone to take refuge under a rocky overhang. As he approached it, he saw there were three almastys sitting under it. Shaghir was a little frightened, but as the rain was by then falling much harder, he decided to stay there anyway, though at a distance from them. They looked at one another. Then, the rain stopped and Shaghir came down to the farm. He did not say anything to anyone.

Very early in the morning, I was awakened by cries, a tremendous noise, and I saw that the shepherds were running to assemble their herds and were taking the cattle down the valley. "Why are they leaving?" I asked. "There are almastys under the rock, up there."... At that moment Shaghir declared: "It's true, there are three almastys sitting up there, I saw them yesterday evening." I was then really angry... I said to Shaghir: "You're an idiot. You were frightened by a bush."

"No," said Shaghir, "I saw them.".
"Well, why didn't you tell anyone?"

"Because the old people have warned: when you see an almasty for the first time, if you tell anyone about it you'll get a bad headache. Well, for me, it was the first time that I have seen one."
I continued not to believe all this. They said to me: "OK, go ahead, go see for yourself."

We were about 10 to 15 people making a half-circle around that rock. We stayed there until dinnertime. Some went away, and others came up. Three almastys were seated under the overhang, two of medium size, and the other bigger. The one which was the biggest was in the middle. They were sitting on rocks, facing us, hunched over, with their heads down. From time to time they raised their heads slightly, and looked at us from under their brows.

Their heads were very ugly, not nice at all. Their faces resembled human faces, but the nose is shorter and flattened. The eyes are slanted and reddish. The cheeks are very prominent, like those of a Mongol or a Korean, but more so. The lips are thin. The lower jaw is receding, as though cut on a bias. The hair is long, like that of a woman, and tangled. The entire body is covered with shaggy hair, resembling that of the buffalo. In some places this is long (torso, chest) and in other places it is shorter (arms, legs).

The big one had the chest of a man. The others had the breasts of a woman, but extremely long and covered with hair. The hair was very dirty. The stink was so strong that we could not stand it. The odour was like that of wild flax, when it grows thickly.

Once, the one seated on the right mumbled something. I did not see their hands clearly, as they were held between their legs. The legs are rather short and bowed. The foot is like that of a man, but more spread out. All were wearing, wrapped around their waists, an old piece of a shepherd's cape. A young shepherd proposed to throw a lasso around one of them and bring it into the village. But all the others cried out that it is forbidden, that they must not be harmed, and that they must not be disturbed. I watched them from a distance of three or four metres, and I even approached to within about one metre. Did I touch them? I should say not! If you touch them, as Allah is my witness, you could no longer eat with your hands afterward, they are so dirty, stinking and repulsive. I remained 1.5 - 2 hours. When I left, other shepherds were arriving. I have heard my father recount that they suckle on cows."

The precise relationship between Modern Man and his Neanderthal cousins has never been properly reconciled. For many years it was thought that Neanderthals had interbred with Cro-Magnon Man to produce our direct ancestors but recent revelations prove that this theory is unlikely. The New York Times Service, on July 11, 1997 reported that:

"DNA extracted from the bones of a Neanderthal man indicates that Neanderthal man did not contribute to the DNA make-up of modern man. While Neanderthal man is classified by palaeo-anthropologists as human, DNA analysis indicates that Neanderthal men never directly contributed to the DNA profile of modern man, and this same DNA evidence also strongly suggests that Neanderthal man never even interbred with modern man. In other words, Neanderthal man contributed nothing to the "gene pool" of modern man."

Although the accounts from elsewhere in the Causacus are ill known, the story of Zana is a well-known one, but there are other reports from elsewhere in the world that appear to deal with very similar creatures. An anonymous article in *The English Mechanic,* 60:429,(1894.), reads:

"At the museum opened at St. Petersburg by Mr. Schultze, a number of remarkable things are to be seen. Krao Farini is among them, the girl who has been nicknamed the "Missing Link," and she belongs to the hairy tribes inhabiting the wild country of Laos, in the northern part of Siam. She is perfectly formed, and although she is not a beauty, her appearance is by no means repulsive.

Her whole body is covered with a thick growth of dark hair. The growth on her cheeks, lips, and chin reminds one of a beard and whiskers; and the nose is also covered with a light down. Still more wonderful to say, she has cheek-pouches, where, like the monkeys, she can stow away tit-bits for future munching, and, like our prototypes, she possesses limbs of singular pliancy. But the most remarkable thing about the freak of Nature is the great wealth of hair on her head. It grows in thick, glistening masses, which fall below the knees. Krao was seen in England some years since, where she very quickly adapted herself to the ways ways of civilization. She can now speak three languages with fluency--English, French, and German.

While Krao was on a visit to Berlin, one of the enterprising Germans proposed to her, but this offer of marriage was not accepted; Krao learned too much independence during her wild life in the woods. Even the celebrated Virchow declared, when he saw Krao, that she was a very extraordinary phenomenon which deserved the fullest attention.

It is interesting that Krao reportedly came from Indo China because it is there that we must go next on our search for our long lost Neanderthal cousins!

The Nguoi Rung, meaning "Forest people", also known as the Vietnamese Wildman is another compelling contender for a surviving Neanderthal.

The descriptions are of a creature approximately six feet tall, almost completely covered with hair except for the knees, the soles of the feet, the hands and the face. The hair ranges in colour from gray to brown to black. The creature walks bipedally and is seen both solitary and with others of its species. Another name for the creature, from the Laotian border, is a local term; Khi-Trau, meaning "buffalo monkey" or "big monkey".

The regions of the Nguoi Rung's apparent habitation are regions where there is little or no human life at all. However the habitat is apparently shrinking, due to bombing during the Vietnam war and following a major deforestation, the Nguoi Rung are running out of places to live.

Since the war, sightings are down, almost to nothing, showing either a marked population of the ape-men were killed during the war, or they were driven to higher mountainous regions where they could hide, without much intervention from any human population.

A map of Indochina showing
Vietnam, Cambodia and Laos

The area known as the "Three borders", where Vietnam, Cambodia and Laos meet, is said to be the centre of Nguoi Rung activity. Reports were so abundant that during the year 1974, during the peak time of the war, General Hoang Minh Thao, the commander of the Northern Forces, requested a scientific survey north of the Kontum region, to attempt to validate the existence of the Nguoi Rung.

No Nguoi Rung was actually found, but the story did not either end, or begin there. Sometime during the 1950's a villager from the Central Highlands went into the forest and vanished without a trace. Three years passed, and the other villagers resigned to the belief that he had became lost and died in the wilderness.

However, one day he returned to the village, his hair had grown long and he was completely nude, and he told them what had happened to him and where he'd been.

When he went into the forest to gather rattan rope he was taken prisoner by animals looking like apes, the animals were bigger than apes, and had extremely long hair, and looked somewhat human in proportion. He was forced to live with one of the female "apes" in a cave deep in the forest.

During the day this female ape kept him in the cave by putting a large stone in front of the entrance. Then she went out to gather food or fish or other edible items.

After more than one growing season, he and the female ape had a female baby, a crossbreed. He didn't explain how the interbreeding came about, or if it had been forced on him or was consensual. Eventually the creatures became comfortable with his presence, and loosened their watch on him, one day he took advantage of their negligence and escaped from the cave, and found his way back to the village.

After telling his story to the villagers, the people in the mountain village often heard long, mournful cries coming from the forest. With the permission of the elders of the village, some of the villagers followed the traces of the man who had returned to the village, back to the cave to where he had lived with the female ape.

When they arrived they found a horrible sight. The entrance to the cave was destroyed and upon entering the cave, they found the baby's body which had been torn apart. The female creature, or any more like it, however, was nowhere to be found.

This story is particularly interesting from a sociological point of view because it is almost an exact opposite to the story of Zana with which we began this chapter. In this case, however, it is the "modern man" who has been captured by the wild people and who has been used as a sexual plaything by them.

This story smacks very much of a piece of folklore - mirroring as it does, so many fairy stories and fables from across the world. If, however, it is true, it suggests that, like Zana and her kin, the Vietnamese wildmen are men rather than animals because of their genetic compatibility with men of our own species.

Ngo Hoang, who during 1950-1952, was an armed agent of the Ministry of Propaganda in the hostile back country of Dac Lac and later was a member of an MIA in the years 1988-1991.

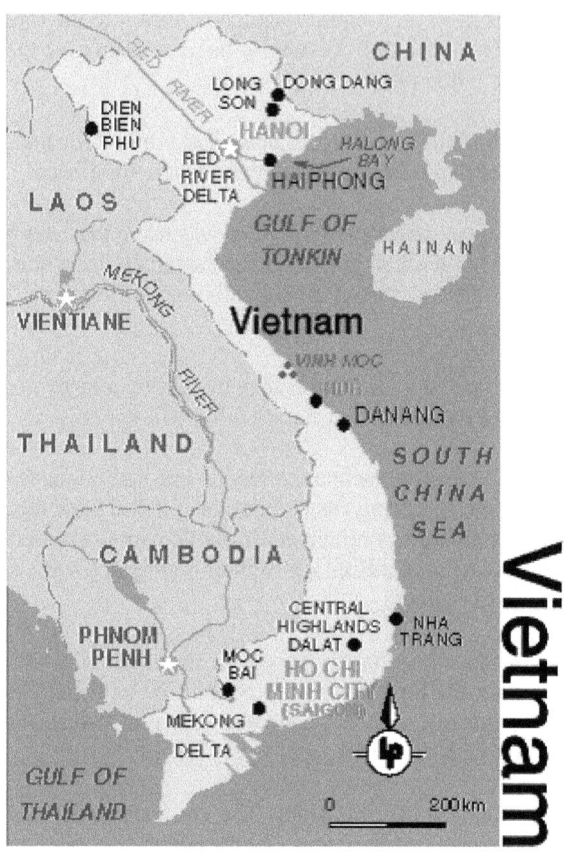

A map of Vietnam showing Da Nang and other places mentioned in the text (Stolen from Vietnam Tourist Board)

As a result he was well acquainted with all the forest areas of Tay Nguyen. He relates the following story: *"We were able to discover these forest men in about 1950 in an area in the vicinity of the Chu Bia mountain chain (which now belongs to the district of Dak Nong - Dak Lak). The footprint this forest man leaves behind is one and one half times as big as that of a normal man (length: about 30 cm, breadth about 20 cm). The big toe is slightly separated from the others. In the middle of the sole of the foot are many folds. After that, some of my comrades of a Pioneer unit saw a big 'man' entirely covered with gray hair. They thought it was an orang utan and wanted to kill him. Fortunately, at that time we had strict orders not to shoot. That is why this man could escape. My comrades told me that he ran very fast, as fast as only a forest creature can run, but no man can run..."*

The Vietnam war - being an occasion when tens of thousands of western soldiers found themselves in terrifying and little-known terrain - has proved to be a rich source of cryptozoological and fortean lore. One of the present authors, in his book *"The Owlman and Others" (1997)* described a sighting of "The Bat Woman of Da Nang" - a terrifying (though sexy) zooform phenomenon, which has parallels in almost every culture across the world. It is unsurprising, therefore to find that American GI's also came back to the land of the free with accounts of apparent man-beasts from the jungles that they were meant to be defending in the name of democracy!

In 1967, at Cam Ranh Bay, a US Army guard detected movement on the perimeter of the camp, the guard fired and whatever had been moving fled the area. When morning came, the area was searched...A footprint and a trail of blood were found. The print looked neither human or ape, a picture was made of the print but no one could decide what to make of the track. The animal was dubbed Powell's ape, after Captain Powell, the depot company commander.

A man named Recom Hiun, who according to our sources is still living in the mountain village of Ky, told of another event. *"The story happened in the mountain village of Ae Thi, M'Drak district (Dak Lak province), probably in the year 1971. One day, Ai Thi tribes people went as usual into the forest to far away streams to fish. They went in a southerly direction to a mountain chain of Cu Yang Sinh, which extends between Dac Lac and Khanh Hoa to the area of Nam Cat Tien and where there is the source of the spring of the Ba River (Phu Yen Province).*

Over a certain stretch of time the Ae Thi could not understand why their traps remained without fish. Then they discovered a strange footprint one and a half times as big as a normal human footprint. They decided to make an ambush so as to capture this creature there and then. Some days passed until one night, before dawn, the saw a strange sight. From out of the interior of the forest came two apes, a bigger one and a somewhat smaller one (soon after it became clear that they were male and female). They were rather slim. They both went into the direction of the ambush. They had long hair on their bodies, including the face, except the eye sockets, mouth, palm and sole. As the male lifted the fish trap and began to tip it over to get the fish, everyone went towards the apes, captured them, tied them up and took them to their village. When a team of scientists in Duc My got news of this event, they went immediately with some South Korean soldiers to visit the village."

Recon Hiun was at this time the director of the Department for the Development of Ethnic Minorities of Khanh Hoa province. He was also there as was another eyewitness. *"The captured ape couple were bound onto two house posts with ropes. The South Korean soldiers then carefully shaved all the hair from their faces and also thoroughly washed both apes. The male was then put by his guards into a striped suit and the female was put into a sarong and Yeng of the Ede women. Both apparently did not know the use of language."*

Mr Recom Hiun said that after that the South Korean soldiers took the captured ape couple directly by airplane to their base at Duc My (Ninh Hoa - Khanh Hoa). Where they were taken from there, nobody knows.

Again this account sounds suspiciously folkloric to us. Contemporary UFOlogy is full of half whispered stories of crashed UFOs which were recovered (sometimes with their inhabitants still intact and alive) from deep in the Vietnamese forests and which were taken to South Vietnamese or US Air Bases where they mysteriously disappeared.

This is part of the mythos of contemporary folklore - that the forces of Uncle Sam are somehow conspiring together to keep various aspects of that shadowy canon known cryptically as "THE TRUTH" from the rest of us.

This has more to do with the innate paranoia of western man at the end of the 20th century than it does with any real interpretation of the facts, but as conscientious researchers we felt that we should report these stories when they come our way.

It was, however, according to many researchers, the American presence in Vietnam which (almost) provided cryptozoology with its greatest ever coup - the discovery of a real, dead specimen of a non-human member of the genus Homo. According to the renowned British zoologist Dr Karl Shuker, writing in his 1993 book *"The Lost Ark - New and Rediscovered animals of the 20th Century"*:

"The case began in the late 1960s, when zoologists Dr Bernard Heuvelmans and Ivan T. Sanderson were informed that a sideshow travelling at that time through Chicago was exhibiting what appeared to be the fresh corpse of a very hirsute man-like entity, preserved within a huge block of ice.

Needless to say, the two scientists were very curious to learn more, and their enquiries took them in mid-December 1968 to a farm near Winona, Minnesota, owned by Frank Hansen who ran the sideshow. He led them to a refrigerated, glass-topped coffin, containing the ice-entombed being that was soon to become known worldwide as the Minnesota iceman."

Hansen allowed the two cryptozoologists unprecedented access to his specimen, and for several days Heuvelmans and Sanderson examined the `creature` as minutely as was possible without defrosting the block of ice in which it had been preserved. Heuvelman's description was exact and thorough:

"The specimen at first looks like a man, or, if you prefer, an adult human being of the male sex, of rather normal height and proportions but excessively hairy. It is entirely covered with very dark brown hair three to four inches long.

Its skin appears wax like, similar in colour to the cadavers of white men not tanned by the sun. The specimen is lying on its back....the left arm is twisted behind the head with the palm of the hand upward. The arm makes a strange curve, as if it were that of a sawdust doll, but this curvature is due to an open fracture midway between the wrist and the elbow where one can distinguish the broken ulna in a gaping wound.

The right arm is twisted and held tightly against the flank, with the hand spread palm down over the right side of the abdomen. Between the ring finger and the medius the penis is visible, lying obliquely on the groin. The testicles are vaguely distinguishable at the juncture of the thighs. "

CFZ YEARBOOK 1999

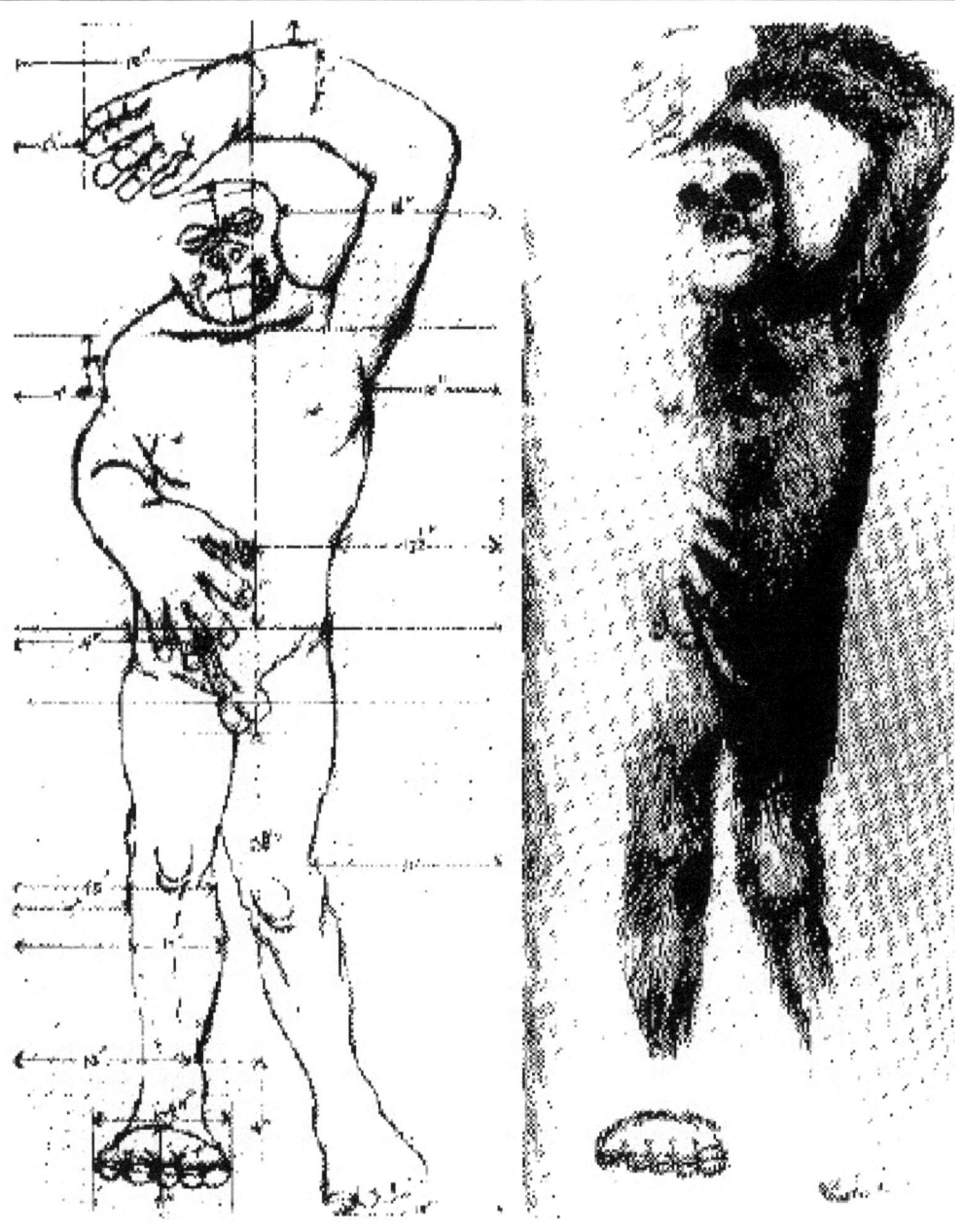

The Best-known image of the so-called Minnesota Iceman

Further, the creature appeared to have been shot in the right eye, the impact apparently blowing out the left eye on impact and apparently blowing out the back of the head as well. A controversy arose soon after, when John Napier, curator of the primate exhibitions at the Smithsonian Institution and another author well renowned within the cryptozoological community also examined the "iceman" but was less impressed. Napier described the creature as some sort of crazy hybrid, apparently combining the worst features of all the primates, and he could not see how something like this could have actually been successful in any environment.

He then tried to obtain permission from Hansen to display it at the Smithsonian, but Hansen told Napier that its owner had retrieved the body, and he (Hansen) would be taking a rubber model of the creature back on tour. Eventually the Smithsonian declared the creature a hoax, but Sanderson died in 1973, still believing the creature to be real. Heuvelmans, to this day contests that the original specimen he and Sanderson examined was real, not a rubber dummy. **

Hansen made several contradictory claims about the provenance of this seemingly unique specimen. He first claimed that it had been found frozen in a block of ice in the Sea of Japan (in a scenario very reminiscent of those claimed for such specimens as the notorious "Feegee Mermaid" by 19th Century entrepreneur and showman Phineas T Barnum). This is palpably untrue, because it is impossible for an animal floating in the sea to freeze in this manner. This is because it would sink and be devoured by scavengers and bacteria before the ice could congeal around it.

Even solid blocks of ice are not total barriers to breakdown by bacteria. The mammoths and woolly rhinoceros found frozen in the Siberian permafrost are actually preserved in frozen mud and slurry rather than in pure ice, which also would have frozen in layers which would have obscured the specimen within rather than the clear block in which the "iceman" is encased which was almost certainly produced by a modern deep freeze.

Hansen then claimed that the creature was a "bigfoot" that had been shot in the wilds of the American heartland. As can be seen from the chapter in this book on Bigfoot and other giant BHMs reported from around the world - whatever the Minnesota Iceman is it ain't bigfoot

---

** This paper was originally written in the summer of 1998. Since then, sadly, Bernard Heuvelmans has died. He died at his home in France aged 84 on August 22nd 2001 (my 42nd birthday). Heuvelmans' archives were given to the University of Lausanne in Switzerland, where they remain today.

He was our first Hon. Consulting Editor, and had been since issue 2 of *Animals & Men*. After his death we decided that we would no longer have this post, and instead created the post of Life-President of the CFZ; a role filled admirably since then by Col. John Blashford-Snell, the semi legendary explorer and soldier.

which is not even remotely human in morphology.

Heuvelmans obviously agreed, because as Karl Shuker (who - like Richard Freeman - appears to be less sceptical on this particular subject than Jonathan Downes) writes:

"Heuvelmans and Sanderson soon published their observations and pictorial evidence in various scientific journals. Heuvelmans was so convinced that the iceman represented a genuine form of hominid wholly distinct from modern man, that in his formal description of this specimen he named its species Homo pongoides *('ape-like man'). He still considers that even if it is not a separate species, it is at the very least a version of Neanderthal man, and should therefore be known as* Homo neanderthalensis pongoide*s"*

An unaccredited Internet article describes the best-known (unless - like at least one of the present authors - you feel that the whole affair is a palpable but rather funny hoax) explanation for the provenance of the specimen:

"With some detective work, Heuvelmans found out the body may have originated in Vietnam. Hansen was a former Air Force Pilot who had connections at Da Nong, not far from a mountainous region which produce [sic] *many wildman reports. At one time, as reported in newspapers in 1966, US Marines were said to have shot and killed a giant ape in the Highlands near Danang, where then Capt. Hansen was stationed. No giant apes have ever been classified in Vietnam, only gibbons, so what was shot was probably a wildman. It appears possible Capt. Hansen obtained the body and arranged to have it flown back in the same manner as the bodies of American soldiers killed in action, refrigerated it and began showing it as a sideshow when he retired. Then, (...) Heuvelmans had given the encased creature the scientific name* Homo pongoides, *Hansen did away with the body in fear of some sort of criminal charges may be brought against him. So, the possible proof of the existence of the wildman may have already been discovered, examined and destroyed, nearly 30 years ago."*

It is undeniable that Vietnam's two most famous exports to the United States during the late 1960's and early 1970's - heroin and dead soldiers - both came back in body bags courtesy of the U.S Army, and it is not inconcievable that if indeed a Vietnamese wildman was shot by the agents of truth, justice and the American way, that its remains were indeed smuggled back to America as if they were several kilos of "Chinese rocks", but the fact remains that Hansen has changed his story repeatedly every few years since Heuvelmans and Sanderson examined the `body`, and has not really shown himself up to be a reliable witness.

Ian Simmons of the *Fortean Times*, a friend and colleague of both the present authors, interviewed Hansen in 1995 whilst he was preparing an exhibition called "Of Monsters and Miracles" for the Croydon Museum, and later wrote up the interview for a controversial piece in FT in which Hansen claims that both Heuvelmans and Sanderson were drunk and abusive as they made only a very cursory examination of the `iceman`. Writing in e-Mail to American Cryptozoologist Loren Coleman several years later, Simmons remembered:

"To me he claimed they arrived at his farm late one evening during heavy snow in a station wagon loaded with photographic gear and asked to see the Iceman. Hansen said it was too

late, but they were welcome to stay until morning and see it then. They then retired to Hansen's basement bar and according to him "Sanderson, despite claiming to be teetotal, put away about half a quart of gin" while Heuvelmans also indulged liberally, all the time leaning on Hansen to let them see the thing that night. Eventually, he said, he gave in, they went out to his cramped display trailer, in the farmyard under snow, hooked into he house electrics to keep he freezer running, and did the examination and took the photos, breaking the thermoglass cover in the process and releasing the smell of decay. Having got the pics they did not wait to stay but careered off into the night with great excitement. I have this recorded on tape from my 95 visit to Hansen, but as I've said before, I'm not sure I believe all Hansen says, he's told various incompatible versions of the tale, is, I suspect, very shrewd and self-serving, and has had nearly 30 years to refine the legend to his advantage and profit. Sanderson had also seen the thing on public display before and after the real/rubber switch and claims there were clear and definable differences. So I have to say I don't know, but err on the side of scepticism until a type specimen is in hand."

Vern Weizel, an Australian Cryptozoologist who fought during the Vietnam War and later married a Vietnamese woman, and who has made a special study of Vietnamese wildmen, later wrote an open letter to Loren Coleman and Ian Simmons which, we think, sums up the whole affair perfectly:

"Dear Loren and Ian,

My opinion is this. Don't be waylayed by the small stuff. Hansen also tried with me to paint Ian as having a hidden agenda of interest. It is irrelevant. I have only one desire and that is to have the specimen made public for examination - not the 'rubber one' but the putative real specimen. The rest only plays into Frank's hands - he will delay as long as he can profit from the information he has (or has fabricated).

The very detailed study by Heuvelmans of the specimen is contrary to his being drunk during the examination (Hansen appears not to have a copy of the book). At least I couldn't do such detailed work while drunk, I don't know about you. And I might say that a description this detailed is sufficient to describe the taxon - though I agree that the lack of a physical sample or of the specimen itself has brought about this whole controversy. Helmut Loofs-Wissowa (with whom I will today communicate this message) is quite incensed that Heuvelmans' efforts are treated dismissively. This has weighed heavily on Heuvelmans who is near despair. People did not take his research seriously.

While I agree with Helmut that this is wrong, dealing with the issue on the personal level commits the same sin. [I have experience with unscientific scientists.] The essential problem remains. Where is the specimen examined by Heuvelmans? And how can its status be verified? Anything else is to continue what is from a scientific point of view a pointless charade. Since I am trapped in Australia, there isn't much I can do. But I really don't agree with any strategy that involves acceding to Frank's greed - or being so unscientific as to extend the breadth of this black comedy to profit from it. Frank Hansen's point of view is that we want to 'get something out of' this specimen: publishing royalties, gate ticket receipts, kudos. I don't care about that. I want the legitimate scientific question solved.

So please don't bother with irrelevancies. We just need the specimen. Frankly I would rather you try to work out a method of examining it in detail. I do not want, even subliminally, to see further oblique publications because to do so will only cement in Hansen's mind the profitability of serialising the 'mystery' that Frank has conjured up. I'm sorry to sound so abrupt, but I am really not amused by Frank's behaviour.

Well, I hope that someone can find a way around this problem.

Cheers, Vern"

Until someone manages to do just that the matter must rest. However the reports of Vietnamese wildmen continued after the war. In the years 1977-78, the biologist Tran Hong Viet, Director of the Department of Zoology at the Pedagogic University of Hanoi No 1, had the opportunity to study the fauna of Tay Nguyen.

In the course of his studies he told local people stories about the forest man. After hearing the stories, these people told him they knew that these creatures existed and they led him directly to a set of footprints which were clearly preserved on a narrow path leading over a mountain, at the base of which was a valley which nobody had ever visited. Mr Viet took a plaster cast of one of the footprints after having photographed it. This cast is still in the collection of the Pedagogic University of Hanoi No 1. The track, which Mr Ngo Hoang described from his experiences in 1950, matched the track in the photo taken by Tran Hong Viet. This led Ngo Hong to make the statement that "In Vietnam there are forest men, there could be no doubt about that."

In 1982, Viet again found and made a cast of a footprint measuring 28 centimetres by 16 centimetres.. The print, while in the range of a human foot in length was much wider than a human foot. The toes were also much longer than that of a human. The print was located on the slopes of the Mom Ray Mountain, near the Cambodian Border in the Kontum Province. The existence of the Nguoi Rung has had an effect on local culture in the Central Highlands, stories of strange gorilla like ape-men living in remote areas of the jungle that hunt humans are imbedded in their lore. The ape-men are said to hunt humans, and upon catching them, hold them captive. The ape-man is said to stand staring at the sky, making sounds like hysterical laughter until nightfall, then the ape-man proceeds to eat the captured man. Apparently, if the ape-men outnumbered the humans, they would attack them and eat them, but if the humans outnumbered the ape-men, the ape-men would run away. The local villagers are so convinced these stories are true, it has become a custom to wear bamboo tubes around their arms so that if they are captured, they can slide their arms from the tubes and escape the ape-men.

Not only have the ape-men tales affected native culture, it has shown up in geographic names for some local regions. There is a jungle in the Central Highlands called "gorilla jungle" named so because workers found footprints like those of a gorilla, as well as the many reports the villagers from this area tell of strange brown-haired animals, like men or apes. Its interesting to note that the villagers living in these remote areas have never heard or seen anything describing part human ape-men from other parts of the world, yet they describe the ape-men very closely to what is said of some of the various ape-men described both in this chapter and throughout the book. No less a personage than Bernard Heuvelmans, the "father of Cryptozoology" who now lives in retirement in France ** described similar creatures in in his classic 1958 book *"On the track of Unknown Animals"*:

"It was Christmas day, in a rubber plantation at Trollak in the south of Perak on the Malay peninsula. A little distance from the other workers, a young Chinese girl of sixteen was tapping a rubber-tree. She was carefully cutting the channel in the bark, down which the gum would run, when she felt a hand laid gently on her bare shoulder. She started and quickly turned her head. And what she saw made her freeze with terror.

A woman stood behind her, if one could call such a hideous and hairy female a woman. She was more like a monkey. Her long black hair hung in a tangle all down her back, in striking contrast to her white skin. She had hair on her arms and on her chest, her eyebrows were large and bushy, and she even had a moustache. She wore nothing but a sort of loincloth of yellow bark, and had a strong animal smell. Her intentions were certainly peaceful. It even seemed as if she had been sent by two no less hairy males, who stood some distance off, as though trying not to frighten the girl. But the female gave a friendly grin, showing long fangs like a beast and gave a sort of croak like a bird. The Chinese girl screamed in terror and fled as fast as her limbs would carry her.

It was not a nightmare. The next day the girl saw the three creatures swimming in a river. A Tamil rubber-tapper, who had also seen the strange trio, confirmed every detail of the girl's account and added that the creatures were of fairly large size and that the males had moustaches which fell down to their waists. When the manager of the plantation heard of this encounter he immediately reported it to the police at Kuala Lumpur, for his people were terrified and did not dare return to work. Two patrols of the Malay security force were sent to the place to hunt the intruders. Corporal Wahab of the Malayan Home Guard saw the three strange creatures on a river-bank, but did not fire on them because he thought that they might be men, such as the Sakai. They dived into the water before his eyes and vanished into the jungle on the far bank. During the next few days the three ape-men, as they were now called, were seen by a good half-dozen people. All their descriptions were the same, and agreed that the creatures were about 5 feet 10 inches high, well built and very hairy all over their bodies. They had been referred to as apes, but they seemed so human in some respects that the police gave the order to bring them back alive.

Local scientific circles began to be excited. Mr G. de G. Sieveking, director of Malayan museums, said that he hoped to organise an expedition to search the Trollak plantations in the hope of tracking down and capturing one of these strange creatures. He pointed out that they wore a strip of skin or bark round their waists and that they ran like men, never using their arms as the anthropoid apes do. In these respects they were human. But their receding foreheads, projecting brow and bushy eyebrows were those of more primitive creatures. These traits and

---

** See footnote on page 28

their long canine teeth, which projected even when their mouths were shut showed, Mr Sieveking cautiously remarked, "that these creatures could be the survivors of one of the first wandering tribes which came to Malaya."

Which wandering tribe? Certainly not the Veddah-like Sakai, nor the pygmy Semang which are too small, dark and hairless. Was he really thinking about a Pithecanthropus? Certainly no human being has canine teeth which project beyond the level of the others. Even in the Pithecanthropus they project only very slightly. But possibly these large canines seemed so strange in a human face that the witnesses exaggerated them. In any case it may have been partly due to a self-inflicted mutilation. The Semang, among others, file their teeth to a point.

Mr Sieveking thought that these hairy men must have already had some contact with civilisation, for they recognised guns and were afraid of them. An inquiry revealed that several of the same creatures had been seen in the states of Kelantan and Perak in 1937. If they had happened to see hunters with guns they would soon realise the danger of firearms."

On the border between China and the former Soviet Republic of Kazakhstan lie the Pamir Mountains from where there have been reports of wildmen from historical times to the present day. Again it is a geographic area that has historically been part of the sphere of Russian influence, and again it is a very wild land about which surprisingly little is known. Most of the best accounts of wildmen from the region come from Soviet Army personnel. This report from Major General Mikhail Topilsky is particularly interesting, as unlike most BHM reports worldwide, this does not refer to a shadowy figure seen fleetingly in the half darkness, but describes, in detail, a REAL man-beast that was killed and examined in minute detail:

"In the autumn of 1925, together with a scouting party we were engaged in tracking down a gang of anti-Soviet guerrillas which was operating in the Western Pamirs. They were trying to shake us off by going to the Sinkiang via the Eastern Pamirs. On our way through the highland villages in the Vanch district we had heard stories about hairy man-beasts, monstrous creatures (I don't remember the local name for them) that lived in the mountains. They were said to be hostile to humans; although they didn't usually attack first.(...)

Once when we were following the gang's tracks along a mountain path and had already reached the permanent snow-line, we saw some tracks running across the path. Our dog took up the scent but refused to follow the tracks. They were very clear and there could be no doubt they were the prints of bare human feet.

They continued for some 150 metres and stopped at the foot of a sheer, barren cliff, which a man could hardly have climbed. Our doctor studied the tracks thoroughly and decided that they were human footprints beyond all doubt (...)

Continuing our chase, we caught up with what was left of the exhausted gang, which had stopped for a rest at a place where the glacier was split apart by a stone cliff. The upper tongue of the glacier hung from the cliff, in which there was a crevice or cave. We surrounded the gang and took up a position above where they were resting. A machine-gun was placed in position. When we threw the first grenade, a man (a Russian officer) ran out onto the glacier and started shouting that the shooting would make the ice cave in and that everyone would be buried. When we demanded that they surrender he asked for time to talk it over with the other guerrillas, and went back into the cave. Soon after, we heard an ominous hissing as the ice began to move. At almost the same moment we heard shots, and not knowing what they meant decided that it was the beginning of an assault.

Pieces of ice and snow started falling down from the cliff, gradually burying the entrance to the cave. When it was almost buried three men managed to escape, and the rest (we learned later that there were five) were buried under the debris. Our shots killed two of the guerrillas and seriously wounded the third.(...)

We questioned him and he gave us the following information. While they were discussing our order to surrender, some hairy, man-like creatures, howling inarticulately, appeared in the cave through a crevice (which possibly led upwards from the cave). There were several of them, and they had staves in their hands. The men tried to shoot their way through. One of the guerrillas was clubbed to death by the creatures. Our narrator received a blow from a staff on his left shoulder as he rushed to the cave entrance with one of the monsters hard on his heels. It ran out of the cave after him, but was shot and buried under a snowslide.

To check up on this strange story we made him show us the exact spot and cleared the snow away. We recovered the body all right. it had been shot three times. Not far off we found a stick made of very hard wood. At first glance I thought the body was that of an ape: it was covered all over with hair. But I knew there were no apes in the Pamirs. Also, the body itself looked very much like that of a man. We tried pulling the hair, to see if it was just a hide used for disguise, but found that it was the creature's own natural hair. We turned the body over several times onto its back and its front, and measured it. Our doctor (who was killed later that year) made a long and thorough inspection of the body, and it was clear that it was not a human being.

The body belonged to a male creature 165-170 centimetres tall, elderly or even old, judging by the greyish colour of the hair in several places. The chest was covered with brownish hair and the belly with greyish hair. The hair was longer but sparser on the chest and close-cropped and thick on the belly. In general the hair was very thick, without any underfur. There was least hair on the buttocks, from which fact our doctor deduced that the creature sat like a human being. There was most hair on the hips. The knees were completely devoid of hair and had callous growths on them. The whole foot including the sole was quite hairless, and was covered by hard brown skin. The shoulders and arms were also covered with hair which got thinner near the hands, and the palms had none at all, but only callous skin. The colour of the face was dark, and the creature had neither beard nor moustache. The back of the head was covered by thick, matted hair. The dead creature lay with its eyes open and its teeth bared. The eyes were dark, and the teeth were large and even and shaped like human teeth. The forehead was slanting and eyebrows were very powerful. The prominent cheekbones made the face resemble the Mongol type of face. The nose was flat, with a deeply sunk bridge. The ears were hairless and looked a little more pointed than a human being's with a longer lobe. The lower jaw was very massive.

The creature had a very powerful broad chest and well developed muscles. We didn't find any important anatomical differences between it and man. The genitalia were like man's. The arms were of normal length, the hands were slightly wider and the feet much wider and shorter than

man's.

A Map of the former Soviet Central Asian Republic of Tajikstan showing the Pamir Mountain Range in the bottom right hand of the country near the border with China

We did not know exactly where we were, because no accurate maps of the Pamirs were then in existence. But we must have been somewhere between the Yazgulem and the Rushan Ranges. As we had completed our task we had to return. (...) The nature of the dead creature presented us with a problem. But it was impossible to take the body with us on difficult trek that lay ahead. Also, it could have caused complications with the local population. We could say, of course, that we were carrying the body of an animal, but the creature looked too much like a human being. We thought about skinning it, but it was too much like skinning a man. In the end we decided to bury the creature where we had found it. We did not try to enter the cave because we were afraid of another cave-in."

Another account of a close encounter with one of the wildmen of the Pamir Mountains comes from the pen of B.M.Zdorik, a geologist, who is a resident of Alma-Ata, Kazakhstan, and who worked in the Pamirs between 1926 and 1938. In a long account, sent to Boris Porshnev in *1959* and published in *The Information Materials of the Snowman Commission,* he wrote:

"In the autumn of 1929, preparing for a hunt, I asked the locals about the fauna in the district. The chairman of the Tutkaul Soviet gave me the following list of local wild animals: wild boar, bear, red wolf, hyena, porcupine, jackal and dev. I was surprised to hear the last name as part of the animal kingdom because according to my previous information `dev` or `div` was a character of Tajik fairy-tales. But here the headman of the locality told me that the dev resembled a thickset man, that it walked on two legs and was covered with brown or black hair. According to the headman, the dev was encountered very rarely in the Sanglakh Mountains, but did turn up now and again either alone or in pairs, male and female. He had never seen young ones, but during the previous summer the Tajiks had caught a grown one alive at a mill on the eastern slopes of the mountain ridge, only a few kilometres from Tutkaul. They kept the dev chained up for two months, feeding it with raw meat and flatcakes of barley flour. Eventually the dev broke its chain and escaped. I did not believe the story and the headman then showed me a villager allegedly injured by a dev. The man had indeed a large scar on the head, but the encounter allegedly so much affected his mind that he was unable to tell me anything intelligible. I then decided that the injury could have been done by a wild boar. Once in 1934, I climbed with much difficulty one of the flat mountaintops in the upper reaches of the Dondushkan. My Tajik guide and I were making our way along a network of narrow paths made by a colony of marmots in the high alpine grass.

Suddenly, a small area opened up in front of us on which the grass was completely flattened, and the ground dug up as if with a spade. On the path were drops of blood and scraps of what looked like marmot fur. And there, right at my very feet, on a heap of freshly dug earth, an unknown creature lay asleep. It was lying fully stretched out on its stomach, about a metre and a half or so in length. I could not see the head and front limbs very well as they were hidden by a bundle of withered grass. I did manage to see the legs and the bare black feet which were too long and too well shaped to be a bear's. But the back was too flat for a bear. The whole body of the animal was covered with shaggy hair looking more like yak's wool than the downy pelt of a bear.

The hair colour was reddish-brown, redder than I have ever seen on a bear. The creature's flanks rose and fell rhythmically as it slept. I stood there frozen with surprise, and at a loss as to what to do. I looked back at my Tajik guide who was following close behind me. He was standing there stupefied, his face as white as a sheet. Then with a gesture he pulled me silently by the sleeve and indicated that we must run at once. Never before had I seen such an expression of terror on a man's face. His fear communicated itself to me, and beside ourselves, without glancing backwards at the creature, we both fled away down the path, enmeshing ourselves and stumbling about in the high grass...

It was only the following day that I learned from the Tajiks, who were rather alarmed by the event, that we had stumbled upon a sleeping dev. They also used some other name for the creature but I don't remember it. According to the inhabitants of the Talbara and Saffedara valleys there were several families of those devs, males, females, and young, living in the mountains. The creatures were considered to be of the animal kingdom, and not supernatural beings, but it was considered to be an evil omen to meet one."

One of the most notable things about these and other accounts from what was once Soviet

Central Asia is that although investigators invariably consider them to be pre-human, the people who live there and who have contact with these creatures on a regular basis consider them to be animals rather than intellectual, biological or social equals.

The reports of Almas or Almasty range across the region from Georgia in the west to Mongolia in the east but always the accounts are remarkably close to each other. According to eyewitnesses the facial features consist of prominent eyebrows, recessed eyes, backward sloping forehead, conical skull, short neck and a very powerful jaw structure. The females of the species have very long breasts that have to be slung over the shoulder when in the running mode. In appearance they closely resemble how we would picture a race of surviving Neanderthals based on our present level of knowledge of their habits and lifestyle.

The International Bigfoot Network describes some of the research that has been done in Mongolia by Professor Rinchen, a member of the Faculty of the University of Ulan-Bator. Once again we are faced with evidence that suggests that whatever they are, the Almasty are close enough to modern man to be genetically compatible.

"In his search for clues to the origins of the Almas, Rinchen came across the product of a human- Almas relationship. A Lama studying at the Lamin-gegen monastery is alleged to have had a father who had been carried away by a group of Almas and during his time with them he was able to father a son by a female Almas. When a passing caravan happened on the scene sometime later, the man and his son managed to escape and return to Mongolian society. The boy proved to be highly intelligent and was so academically brilliant that he had no trouble in being accepted by the lamasery where he became a noted scholar."

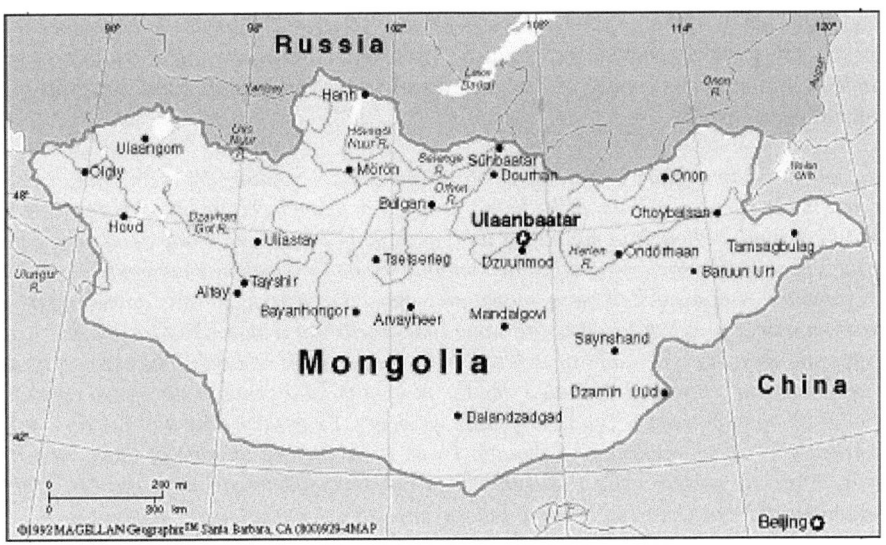

Mongolia

One of the biggest problems when researching fortean phenomena especially at second hand is that when one is faced with a statement like the one pointed immediately above it is almost impossible to check it. Indeed, on a number of occasions during our researches we have found that a number of reputable and well respected sources have quoted, and re-quoted a piece of information that was not only erroneous in the first place but was credited to someone who never actually existed.

This quasi-academic extension of the old parlour game of `Chinese Whispers` is often quite innocently done, and we are sure that we have been unwittingly guilty of this a number of times during our careers. Whenever possible, therefore, we do our best to verify that the people we have quoted are, indeed, bona fide sources. Our attempts to contact Professor Rinchen in order to get him to confirm this story, however, were fruitless.

We did, however, manage to confirm that Professor Rinchen Barbold is a well-known member of the faculty and an expert on vertebrate palaeontology. He has written widely on subjects compatible with a deep interest in cryptozoology, and for the moment, therefore, until we are faced with any evidence to the contrary, we shall let is evidence stand. Another tantalising story from Mongolia, and this time one whose provenance is slightly easier to verify comes from Mr. Damdin of the State Museum of Ulan Bator who spent several months in the Khovd and Bayanolgy provinces, wrote:

"It was happened at about ten o'clock of 26th June, 1953. I remembered the time, day and month because this day had utterly surprised myself and was engraved on my heart. At dawn of that day I went to search my lost camels in the direction of the so called Red Mountain of the Almases, It was a beautiful sunny morning when I dropped into the ravines, The wind spread a fragrance of highland flowers and herbs but I was in a hurry to leave before a midday heat this labyrinth of canyons and ravines. My camel climber up and down in craggy defile. Suddenly I saw in the corner of secluded ravine under two small ammodendron bushes something of camel-colour.

I approached and saw a hairy corpse of a robust humanlike creature dried and half buried by sand. I had never seen such humanlike being covered by camel-colour brownish-yellow short hairs and I recoiled, although in my native land of Sinking I had seen many dead men killed in battle. But who was this strange dead thing - man or beast? I decided to return back and thoroughly examine it. I approached once more and looked down from my camel. The dead thing was not a bear or an ape and at the same time it was not a man like Mongol and Kazakh or Chinese and Russian. The hairs of its head were longer than its body. The skin on its groin and armpits was darkened and shriveled like the hide of a dead camel. I have also examined a terrain near its body and never found any rests or wears. Fear seized my heart. I remembered the old tales of Vetala-Vampires and thought I was to see before me one of them. And I hurried away. After my return home I had informed our local administration and Mr. Chimeddorje, manager of Fruit Growing Station, but anyone gave attention to my account."

Many researchers including the present authors believe that the elusive Almasty of Mongolia represent some of the best chances that contemporary scientists will ever have of seeing a living Neanderthal. After all, the sightings, which have continued to the present day, have been occurring for a very long time. The earliest known printed reference to the Almas was by Hans Schiltberger, from Bavaria. In the 1420's he travelled through the Tien Shan mountains as a prisoner to the Mongols, in a journal he kept he wrote:

"In the mountains themselves live a wild people, who have nothing in common with other human beings, a pelt covers the entire body of these creatures. Only the hands and face are free of hair. They run around in the hills like animals and eat foliage and grass and whatever else they can find. The Lord of the Territory made Egidi a present of a couple of forest people, a man and a woman, together with three untamed horses the sizes of asses and all sorts of other animals which are not found in German lands and which I cannot therefore put a name to."

With the new political openness across the former Soviet Union and its allies, western expeditions are now able to enter areas that have been forbidden to their forebears. In Mongolia alone, for example, according to the August 1998 edition of *Fortean Times* there were no less than four expeditions searching for various zoological anomalies including the fabled `Death Worm`. **

Surely it is only a matter of time before they return with conclusive evidence that our genetic cousins still roam the wilder places of our planet?

Another approach towards solving the riddle of the Eurasian wildmen is to approach them from a socio-cultural perspective. Many cryptozoologists have drawn a link between the ancient mythological archetype of the satyr and putative neanderthal survivors. Michel Raynal wrote, for example:

"Half humanoid, half goat, the Satyr is the mythological prototype of the wildman : hairy, living in mountains, fond of wine and girls... But it has been suggested by Boris F. Porshnev and Bernard Heuvelmans that Satyrs are partly based upon relict Neanderthal Men, as demonstrated by their features : hairyness, the "goat's foot" (which in fact means a foot adapted to climbing, a mountaineer foot, typical of modern wildmen and of the most specialized fossil Neanderthals)".

He goes on to draw links between depictions of satyrs from classical Greek remains in the Crimea (geographically a mere stone's throw from the site of contemporary reports from Georgia and the Transcausacus, and the facial reconstructions extrapolated from Heuvelmans's depiction of The Minnesota Iceman. (a.k.a *Homo pongoides*). As we have shown earlier in this

---

** The CFZ didn't start their self-funded programme of foreign expeditions until 2003, but our third trip (in 2005) was to Mongolia in search of the deathworm. The full account of the expedition can be found on the CFZ website.

Our friend and colleague Adam Davies has carried out several expeditions to Mongolia, some in conjunction with American TV channels, and the CFZ is planning a return visit in 2010 or 2011.

chapter, we think that it best to draw a discreet curtain across this particular episode in cryptozoology's history, but the basic concept is an appealing one.

Our old friend and colleague Dr Karl Shuker has also theorised about the link between putative Neanderthal survival and the Eurasian myths of lecherous wildmen. Writing in *"The Unexplained" (Carleton 1995)* he notes:

"In Greek mythology satyrs were semi-humans with hairy legs, hooves, tail and short horns of goats: But did they have a basis in reality? This unexpected prospect was raised in a stimulating paper published in the scientific journal `Human Evolution` in 1994 by Dr Helmut Loofs-Wissowa from the Australian National University's Faculty of Asian Studies. In ancient classical art satyrs were frequently portrayed with a prominently erect penis even when engaged in non-sexual activity. Indeed, it was this characteristic that earned them their reputation for sexual licentiousness. However, Dr Loofs-Wissowa believes that all this is fallacious and in fact the satyrs were displaying a physiological condition known as `penis rectus` in which the penis assumes a horizontal position even when flaccid. Among modern humans this condition is only recorded amongst the Bushmen of South Africa, but is often portrayed in prehistoric cave art including some Upper Palaeolithic examples from Europe in which the figures exhibiting penis rectus condition are hairy humanoids".

As Shuker points out, this is quite good supportive evidence for the continuing existence of Neanderthals at least into the times of the Classical Greeks. Other writers including Francis Hitching (in *"The World Atlas of Mysteries"*) had also drawn the conclusions that the classical satyrs were in fact surviving Neanderthals so this seemed a particularly interesting line of enquiry.

Loren Coleman and Patrick Huyghe - the authors of*: "The Field Guide to Bigfoot, Yeti, and Other Mystery Primates Worldwide"* (New York: Avon Books, 1999) disagree and believe that the available evidence leads towards the creation of yet another putative mystery hominoid:

"The Erectus Hominid is probably the least known of the world's mystery hominids. The reason for this is simple: most of the beings in this class have in the past been misidentified as Neanderthal. The Erectus Hominid is human-sized to about six feet tall. Its body is also within the standard human range with a slight barrelling of the chest. They are partially to fully hairy, with head hair longer than their body hair. The males of the class normally display a semi-erect penis."

The final stop on our journey to find our closest relatives is even less well known than the rest of the topics we have discussed in this chapter. We must go to, of all places, Greenland, but it is here that the final evidence for the survival of other members of the genus *Homo* may have been found over half a century ago. Legendary Fortean Mark Hall writes:
"More than eight hundred years ago the Norse from Iceland and Norway inhabited parts of Greenland. In 1126 they established the community of Gardar (today called Igaiko) near the head of the Igaliko Fjord. It was established as the Episcopal seat for all of the Greenland colonies. The bishops lived at Gardar.

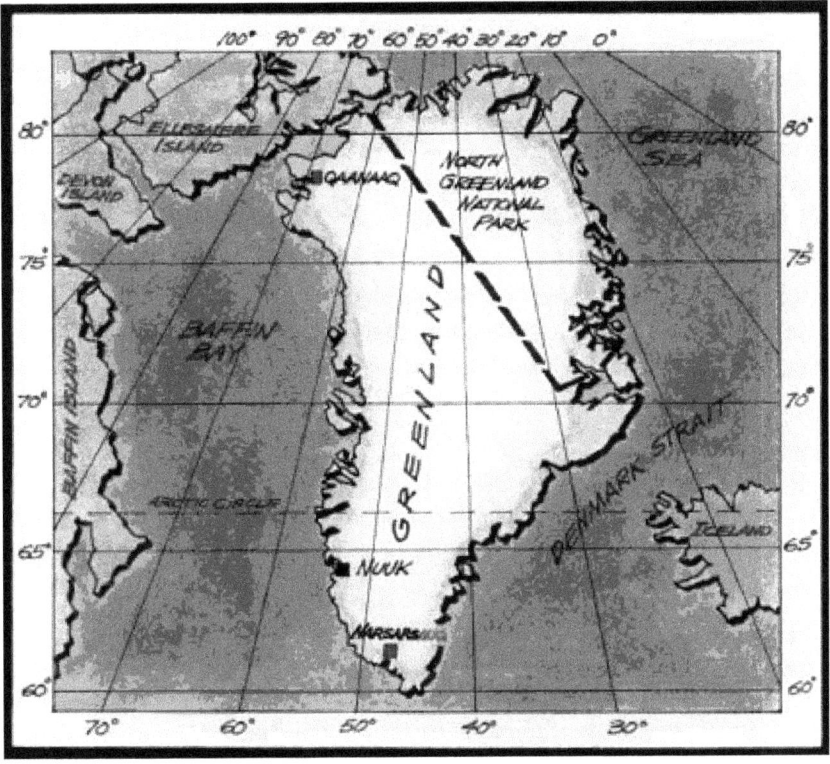

In 1926 Danish excavators from the National Museum of Copenhagen located the archaeological remains of Gardar. During the excavation they came upon the skeletal remains of Bishop Jon. He was buried wearing an Episcopal ring and holding a bishop's staff. The other burials beside the now-vanished church included an unusual skull. The man who studied the skull in Denmark, Prof. F.C.C. Hansen, was an expert in the study of the Eskimos. He did not think it was related at all to them He had to conclude that it came from a Norseman who had somehow reverted to an earlier stage of human evolution. The idea of such a sudden reversion to a primitive state has no standing in physical anthropology today. Another anthropologist thought the skull resulted from a severe case of acromegaly, a clinical problem causing an enlarged deformation of the bones. Prof. Hansen replied that such a problem did not agree with "the regular proportions seen in the jaw and cranial bones of the Gardarene skull and the harmony of all of its features. What Prof. Hansen definitely saw in both the cranium and the skull were "in an extreme degree characters which mark the human skulls of an ancient type, such as the La Chapelle and Rhodesian skulls." People have asked for decades, "If there

exist non-human higher primates in North America where are the bones?" Here in the skull of Homo gardarensis we have the anomalous evidence that people have been talking about in the last half of the twentieth century. Other bones have been found but misplaced by the scientists entrusted with them. It is a matter of record that a calverium found in California in 1965 has been lost. It is out of reach in a poorly marked crate amid a myriad of crates in the warehouse of the museum at the University of California at Los Angeles. A second case of loss occurred when unusual skulls turned up in northern Minnesota. They were found in the Boundary Waters Canoe Area in 1968. By the time the discovery was discussed publicly the bones had been sent to the Smithsonian Institution and lost there."

This is interesting and very reminiscent of the first discovery of Neanderthal remains as recounted by Ian Tattershall (1995)

"Mayer's examination of the bones from the Feldhofer cave suggested several things to him. He noted, for example, that the thigh bones and the upper front part of the pelvis were somewhat curved, as in lifelong horsemen. These characteristics, he claimed, might also have been exaggerated by childhood rickets, a vitamin deficiency disease. The left arm had been fractured and had healed badly; and Mayer claimed that this injury was the key to the unusual shape of the skull: it was the constant frown brought on by the pain of the injury that had caused the formation of the bony ridges above the eyes! Putting all the evidence together, Mayer proposed that the remains were those of an unfortunate deserter from the Cossack cavalry that has paused near the Rhine in January of 1814, before proceeding onward to attack France".

Here, however the evidence for *Homo gadarensis* being something innately far more interesting than the malformed remains of a freak of nature is far more compelling. As Hall writes:

"The find of Homo gardarensis was the subject of short and inconclusive debate and then forgotten. The context to explain the strikingly primitive and well-proportioned features of the skull was known even in 1926. The Eskimos in that part of North America described the beings who preceded them. These beings were known to the Eskimos as the Tornit (also Tunnit; singular Tuneq or Tunek). These Tornit tried to live in the same way as the Eskimos but were not as efficient. This was so despite being larger and far stronger than the Eskimos. The Eskimos had come into conflict with the Tornit and methodically disposed of as many of them as they could. Same were still living said the Eskimos in this century, in remote parts of the Far North. Archaeologists are now certain that those people whom they call the Dorset Culture represent the physical remnants of the Tornit. They are far less certain about who the culture hearers were and why they disappeared. Here I am presenting the answer that these culture hearers were the Taller-hominids who have not entirely disappeared.

Both the Eskimos and the Tornit hunted the same animals. They possessed different tools and used different structures as homes. The Dorset people are considered to have lived in the Canadian Arctic from two thousand five hundred years ago until at least one thousand years ago. In the archaeological record they were overtaken by the Thule tradition that developed around the Bering Sea and spread across the North. The Thule tradition was then modified to become the historic Eskimos we know today. Archaeologists don't know how the Dorset disap-

peared but the Eskimos have a lot to tell us about that. We will turn to older records of the Tornit after noting how archaeological research has tended to support the Eskimo accounts".

A modern anthropologist, Moreau S. Maxwell, wrote at length about the Dorset people and described a culture which, whilst having reached a surprising degree of sophistication, does not seem to have much to do with that of the present day Eskimo people:

"Some of these legends about the Tunnit derive authenticity from archaeological evidence. They are said to be dwarfs, or giants, with prodigious strength, and there are a few Dorset tent rings that bear this out. I once excavated such a site on an outwash plain of northern Ellesmere Island. The tent ring, which contained Dorset stone tools, was barely 2 min diameter, yet the huge boulders that formed the walls ranged from 100 to 200 kg. Each of these rocks had been dragged from several hundred meters away.... According to the stories, the Tunnit, since they had no sled dogs, used this prodigious strength to drag killed walrus home using only a small single-man sled. (The archaeological evidence is that among the Dorset, dogs were very scarce and only sail hand-drawn sleds were used.)"

According To Mark Hall and others, The best information about the Tornit comes from Ernest William Hawkes (1883-?) who preserved the following description when writing *The Labrador Eskimo* published in 1916. Again, the overwhelming impression one gains from Hawkes' writings is that the Tunnit were a very real (but very different) people to the Eskimo or Innuit:

"Considerable information regarding the Tunnit was gathered on this trip. It is placed in the mythological section for convenience m comparing with the traditions gathered by other writers. The author is of the opinion that the Tunnit are entitled to an historical position in northern Labrador.

Tunnit (Tornit, Baffin Island), according to tradition, were a gigantic race formerly inhabiting the northeastern coast of Labrador Hudson Strait, and southern Baffin Island. Ruins of old stone houses and graves, which are ascribed to them by the present Eskimo, are found throughout this entire section, penetrating only slightly, however, into Ungava Bay. Briefly we may say that there is evidence, archaeological as well as traditional, that the Tunnit formerly inhabited both sides of Hudson Strait. The oldest Eskimo of northern Labrador still point out these ruins, and relate traditions of their having lived together until the Tunnit were finally exterminated or driven out by the present Eskimo.

According to the account given by an old Nachvak Eskimo, the Tunnit in ancient times had two villages in Nachvak Bay. Their homes were built on an exposed shore (the present Eskimo always seek a sheltered beach for their villages, where they can land in their kayaks), showing that they had little knowledge of the use of boats. When they wanted boats, they stole them from the Eskimo. From this thieving of kayaks the original quarrel is said to have begun. For all their bigness and strength, the Tunnit were a stupid, slow-going race (according to the Eskimo version), and fell an easy'! prey to the Eskimo, who used to stalk them and hunt them down like game. They did not dare to attack them openly, so cut them of one by one, by following them, and attacking and killing them when asleep. Their favourite method was to bore holes in the foreheads of the Tunnit with an awl. Two brothers especially distinguished them-

selves in this warfare and did not desist until the last of the Tunnit was exterminated. The Tunnit built their houses of heavy rocks, which no Eskimo could lift. They used the rocks for walls, and whale ribs and shoulder blades for the roof. At the entrance of the hose two whale jawbones were placed. Ruins of these houses can still he seen, overgrown with grass, with the roof fallen ~ They may he distinguished from old Eskimo igloos by the small, square space they occupy. The Tunnit did not use the bow and arrow, but flint-headed lances and harpoons with bone or ivory heads. They were so strong that one of them could hold a walrus as easily as an Eskimo a seal They did not understand the dressing of seal-skins, but left them in the sea where the little sea-worms (?) cleaned off the fat in a short time. The Tunnit dressed in winter in untanned deerskins. They were accustomed to carry pieces of meat around with them, between their clothing and their body, until it was putrid, when they ate it.

The Tunnit were very skilful with the lance, which they threw, sitting down and aiming at the object by resting the shaft on the boot. For throwing at a distance they used the throw-stick. They did not hunt deer like the Eskimo, but erected long lines of stone "men" in a valley through which the deer passed. The deer would pass between the lines of stones and the hunters hidden behind them would lance them. Remains of these lines of rocks may still he seen Their weapons were much larger, but not so well made as those of the Eskimo, as can he seen from the remains on their graves. The men used flint for the harpoon heads, and crystal for their drills. The women used a rounded piece of slate without a handle for a knife. They used a very small lamp for heating purposes, which they carried about them. For cooking they had a much larger lamp than the Eskimo. Until trouble arose between them, the Tunnit and the Eskimo used to intermarry but after it was found than an alien wife would betray her husband to her people, no more were taken. A Tuneq woman, who betrayed the Eskimo of the village she lived in to the Tunnit, had her arms cut off.

After than no women were taken on either side....

The Tunnit were gradually exterminated by the Eskimo until only a scattered one remained here and there in their villages. How thorn these were overcome by stratagems is handed down in the tales of the giant at Hebron, said to he the last of the Tunnit, and Adlasuq and the Giant. The giant allows himself to he bound in a snowhouse, and is slain by the Eskimo hunters.

The story has attained a mythological character in Baffin Island but is ascribed by the Labrador Eskimo directly to the Tunnit A story about the Tunnit, giving considerable circumstantial detail, was obtained from a Nachvak woman:

"At Nachvak the Tunnit were chasing a big whale (this was before the time of the present Eskimo). They were in two skin boats abbot twenty men and women in each boat. They had the whale harpooned, and were being towed round and round the bay by him. Somehow the line got tangled in one of the boats and capsized. The other boot with the line still made fast to the whale went to pick up the people in the water, and was capsized too. Another boat came off the shore, and picked up some of the people in the water. Most of them were drowned.

They were buried under a hill on a big bank near Nachvak. There are some thirty graves on this bank, with pots, "harpoons, and knives buried by the graves. Even the remains of the boats are there. The 'wives and pots are of stone. The harpoon blades are of flint. The Amax were much larger than the present boats." My informant added that there were also remains of bows and arrows. "The bows were of whalebone and the arrows of flint.?"

Further information was obtained from another informant.

<u>Tunnit Houses.</u>

The houses had long stone passages. The two posts at the entrance were of whale jaw-bones and shoulder blades on top. The walls were of stone and turf The roof was formed of whale ribs on props, and covered with turf. The roofs of the houses have now fallen in, but the walls are still intact.

<u>Tunnit Boots.</u>

The Tunnit did not know how to manufacture waterproof boots. They took a long strip of seal-skin with the hair on, and wrapped it around the feet, starting at the toes. For a sole they would a flat, square piece of skin, cut holes around the edge, "reave" it up with a drawstring, and tie it around the ankles."

There seems very litle doubt that these enigmatic inhabitants of the frozen wastes lived in Greenland until a very recent time. Two questions remain... `Who (or what) were they? ` And `Are they still there? `

Although as we - and others cited above - have shown, the Tunnit seem to have achieved a far different culture than the Eskimo people, they do not seem to be Neanderthals either. They are too tall, for one thing, and also the level of cultural sophistication that they have reached seems to be beyond that ever achieved by *H.s.neanderthalis*. Our best guess is that they were a very primitive race of people, probably kin to the Mesolithic hunter-gathers whom we suspect still live in small groups in the Transcausacus, who kept their hairy skin and evolved a large size in order to cope with the climate of the frozen north!

However, the morphological differences between *H.gardarensis* (and in the abscence of any further information we must, we think, assume that the `Tunnit` are indeed synonymous with *H.gardarensis*), and other known races of modern man, are such to suggest that further work needs to be done in this little known area of palaeoanthropology.

We started this chapter with the United Nations Declaration of Human Rights, and it seems fitting that we should end on another socio-philosophical point. Throughout this chapter we have been discussing human evolution (or the lack of it) as if it were a one-way process. Imagine, however a scenario that is seldom discussed because of the unfortunate political and socio-cultural connotations that it has. If a culture, or a species can evolve, then surely it can also de-evolve or degenerate.

The chief objection raised in the "almas as Neanderthal" scenario is the ultra-primitive nature of the creatures themselves. As we have shown above, it seems from recent archaeological discoveries that neanderthals had mastered fire, built workable tools and weapons, lived in co-operative communities and even seemed to have a quasi-religious concept of an afterlife. At a site in Le Moustire in southern France, the remains of a 16-year-old boy were unearthed. In his hands was an excellently made stone axe, beneath his head a pile of flints, and scattered about him were cattle bones. It seems highly unlikely that such useful items would have been wasted, but they may have had a religious significance. Tools and food for his journey perhaps? Another grave in Uzbekistan, in what was formerly Soviet Central Asia had an early headstone made from six pairs of ibex horns.

In a stark contrast, the almas, almasty and similar creatures reported from across the world, do not seem to have any culture whatsoever. Their tool-use extends only to hurling rocks and swinging clubs; the only clothes associated with them are ones pilfered from modern humans. They make no artefacts of their own, and only use natural shelters such as caves.

If the creatures we have described are indeed surviving Neanderthals, what could have caused such a total cultural degradation? If - as postulated - indirect competition with Cro-Magnon man pushed Neanderthals into marginal areas, the limited resources would not have been able to support large tribes. As tribal groups broke up and links were lost Neanderthal society would have gradually fallen apart. Vital knowledge such as the construction of spears and the making of fire could not have been passed down from generation to generation as easily as before. Religious ceremonies would have also been lost as shamans died without successors,.

Throughout human history we see what happens when a human society comes into contact with one of greater technological ability. The weaker and more primitive society soon disintegrates and its people lapse into obscurity and usually disappear. Examples from recent human history include the disappearance of the `Indians` of Tierra Del Fuego, the original inhabitants of Easter Island, and - possibly most tragically - the decimation of the Native American Tribes of what is now the United States.

It seems likely therefore that Neanderthal culture went out with a whimper rather than a bang, their society slipping away by erosion rather than in a conscious act of Cro-Magnon genocide.

It seems however, that the accepted scenario whereby as the UNESCO Declaration states *"All human beings belong to a single species and are descended from a common stock"* is, as we stated at the beginning of this chapter, palpably untrue. We are most certainly NOT alone. However it is facile merely to state that there are surviving groups of *Homo sapiens neanderthalensis* living in the wilder parts of Eurasia. The truth is as Oscar Wilde so sagely remarked is never pure and seldom simple.

It seems from our researches that there are several different groups of primitive people living in the wilder parts of our planet.

* MESOLITHIC ERA HUNTER GATHERERS. Although it is common knowledge that `stone age` era people still live in southeast Asia, parts of Africa and the Amazon basin, it seems almost certain that there are small family groups of primitive men and women living in

Georgia and the Transcausacus. They may also have lived until recently (and may still live) in the remote parts of Greenland. These people may be primitive, and may have morphological differences (particularly hirsuteness) from modern man, but they are still of the same species as us! The appellation of *Homo gardarensis* would seem, therefore to be unsound.

* THE ERECTUS HOMINOID. Whether or not one is entitled to use the nomenclature of *Homo pongoides* which was coined in the late 60's by Heuvelmans and Sanderson for the `Minnesota Iceman`, it seems that the description they gave, (which has been used most recently by Loren Coleman and Patrick Huyghe) is apposite for the hominoids currently reported from parts of mainland south eastern Asia. At present their precise position within the genus Homo is not known, and until a specimen is finally secured it seems appropriate to use Heuvelmans and Sanderson`s nomenclature.

* "TRUE" NEANDERTHALS. The only `true` specimens of *H.s.neanderthalensis* seem to occupy vast swathes of the former Soviet Central Asia where they have been safe from discovery by scientific orthodoxy because of the political waste ground in which - by no fault of their own - they have found themselves living.

Zoologists are not meant to be emotive. The `High Priests` of contemporary scientific orthodoxy frown upon flights of fancy. However, as our investigation into the putative survival of our closest relatives draws to a close perhaps you will allow us to let our imaginations run riot. As we write this, a great sadness comes over us.

One wonders what the almas makes of this changed world in which we all now live. Are they lost in a forest of straight lines? Do they dream dreams haunted by racial memories of woolly rhino, mammoths and other long vanished beasts of the ice age? As we, their cousins lay waste to the planet; they cling, ignorant to a bleak existence in the worlds remotest regions awaiting their inevitable and long overdue extinction.

Perhaps we are wrong. Just perhaps enough of them survive to await not extinction but inheritance. When the planet's current masters finally commit the racial suicide that we seem so intent upon, perhaps our "older brothers" will be waiting in the wings to reclaim their birthright.

PELORUS JACK

by

JAMES COWAN

PREFACE.

So many enquiries still reach New Zealand about "Pelorus Jack," the solitary white dolphin which once frequented the waters near the French Pass, on the northern coast of the South Island, and so much curiosity is displayed by travellers about this "great fish story" as some call it, that this little book has been reprinted from the original work (1911) and revised in order to place on record some facts and folklore about the famous creature of the sea. The story of "Pelorus Jack," and his lonely life, and his habit of meeting coastal steamers plying between the two main islands of New Zealand, has often been treated as a mere "travellers' tale" But "Jack" was no myth, whatever one may think of the curious Maori tales that cluster round him and some of which are given in this book. A sketch-map showing "Jack's" sea-habitat is given, and at the end of the book will be found the Government Order-in-Council (not an Act of Parliament) protecting all fish or mammals of his species in the waters of Cook Strait. Amongst the illustrations is a photo of Kipa Hemi Whiro, the last of the tohungas or wise men of the ancient Ngati-Kuia tribe, who gave the writer the poetic legends of "Pelorus Jack"' - or "Kaikai-a-waro" as the Maoris called him.

- *J.C. Wellington, New Zealand, 1930.*

THE STORY OF
PELORUS JACK:

THE WHITE DOLPHIN
OF FRENCH PASS, NEW ZEALAND

WITH MAORI LEGENDS

BY

JAMES COWAN

SECOND EDITION

WHITCOMBE & TOMBS LIMITED

AUCKLAND CHRISTCHURCH DUNEDIN WELLINGTON
MELBOURNE SYDNEY LONDON

THE WHITE DOLPHIN OF PELORUS

Ever since the dim old Greek days when "the Nereids danced, the Sirens faintly sang," and when a dolphin succoured the Lesbian minstrel Arion and bore him safely to the Corinthian shore, the sociable dolphin has been regarded as a friend of seafarers. It was sacred to Dionysus and Apollo, just as the famous lone cetacean of French Pass was *tapu* to Tangaroa, the Maori-Polynesian Neptune. The dolphin seems to have been looked upon as a saviour of life, and a rescuer of the drowning; and when old sailors died at sea those that did not become albatrosses turned into dolphins. The "wind-jammer" sailor of today has a sprightly affection for these ship-loving ocean-creatures.

Maori folklore abounds with references to *taniwhas* of the deep, sea-deities which were originally men, and which exercised benevolent and maleficent powers over humans who ventured out upon the deep. Some of the heroes of old became *maraki-hau* or mermen; and on many of the carved slabs which decorate the walls of Maori tribal meeting-houses to-day you may see these *maraki-hau* chiselled, with strangely-shapen mouths scooping in the fish of the sea, and their Triton tails curved like a dolphin's or coiled in snaky spirals like a sea-horse's. Most famous of all these Maori *taniwha* fish and *maraki-hau* was "Kaikai-a-waro," which the *pakeha* called "Pelorus Jack."

As Triton with his "wreathed horn" preceded his ocean-riding father Poseidon, so the scythe-finned *taniwha* of the Maori seas escorted their chiefs' canoes; and so did "Kaikai-a-waro," playing swiftly around the bows of the Trans-Cook Strait fire-canoes, as if leading them on their way, a wonder and a delight to thousands of sea-travellers. "Pelorus Jack," the pilot-dolphin of the French Pass and its vicinity, was probably the most notable creature of his kind in the world. His remarkable habit, continued through more than twenty years, of daily and nightly meeting steamers passing through a certain part of Cook Strait, dividing the two principal islands of New Zealand, and of accompanying them on their way for several miles, and his uncanny, almost human sociability, invested him with extraordinary interest.

Many a visitor to New Zealand made a trip across the Strait from Wellington City to Nelson for the express purpose of seeing - and in many cases photographing - "Pelorus Jack," playing like a sea-god at the steamer's bows. With some of us, perhaps, there was a feeling akin to the old Maori belief that the lone dolphin of Pelorus had something of the supernatural about him. Certainly he was no ordinary creature of the sea. Why did he live alone? Why did he spend his days and nights - or a good part of them - darting ahead of and around the steamers that pass the head of Pelorus Sound? Did some mysterious ocean wireless give him notice of their coming? Why was his ocean beat always the same? Those were some of the questions everyone

---

EDITOR'S NOTE: My first introduction to the story of Pelorus Jack came in a brief mention in a book which has been one of my sacred texts for forty years now—Arthur Ransome's *Swallows and Amazons*. We received a photocopy of the booklet about Pelorus Jack from a colleague in Australia (I think it was Tony Healey, but after nearly a decade my memory has given out on me), and I was so pleased to be able to read this obscure pamphlet for myself that I decided that it should be included in the CFZ yearbook.

asked who had seen or heard of "Pelorus Jack."

All sorts of romantic yarns have been written by imaginative travellers about this strange sea-creature. Some people even went to the trouble of inventing Maori legends of their own about

him. But there was no need to do so. If they only knew it, a wealth of *bona-fide* Maori folk-lore surrounds old "Jack." The haunts of this remarkable fish - he was not really a fish, but it is convenient to call him one - were the waters about the entrance to Pelorus Sound, a long and winding sea-fiord which runs far into the mountainous northern part of the South Island. A labyrinthine coast-line this, of capes and headlands, crooking their arms round lake-like bays and long quiet salt-water reaches; bays within bays; and an archipelago of rocky islets sprinkled about the surf-beaten outer coast. Pelorus Sound and Queen Charlotte Sound have between them a sheltered inner coastline of something like five hundred miles; Pelorus alone has a coastline of three hundred miles. For days you may cruise on an even keel along these sweet waters serpentining between the steep blue ranges. Magnificent harbours here, locked in by the highlands of Marlborough and Nelson. But the funnel-like Strait is oftentimes a stormy place; and it was this stormy stretch of sea - or a portion of it - that "Jack" patrolled for many a year, and if we are to believe the Maori tales, for many generations.

"Pelorus Jack's" sex was uncertain, but everyone seemed to take it for granted that it was a "he." So "he" let it be. He was a dolphin of a bluish-white colour, tinged with purple and yellow, and with irregular brown-edges scratch-like line covering the upper surface of his body.

His flippers were dark in hue, mottled with grey. He was about fourteen feet in length - as nearly as could be judged - for he didn't stay still very long - and he was blunt of nose, humped of forehead, with a high falcate or scythe-shaped dorsal fin and a narrow fluked tail.

So much for "Jack's" personal appearance. He was scientifically classified in 1904 as Risso's dolphin *(Grampus griseus);* and accordingly all cetaceans of that species in Cook Strait were proclaimed as protected under an Order-in-Council of the New Zealand Government, with the object of preserving "Jack" from the gun or the harpoon of life-destroying humans. For a long time "Jack's" actual species was in doubt. Sometimes he was described as a white whale, sometimes as a white shark. But there was nothing of the shark about the blithesome dolphin of Pelorus-mouth.

> ISLINGTON, Governor.
> ORDER IN COUNCIL.
> At the Government Buildings, at Wellington, this twenty fourth day of April, 1911.
> Present:
> THE HONOURABLE JAMES CARROLL PRESIDING IN COUNCIL.
>
> WHEREAS by Order in Council dated the twenty-third day of May, one thousand nine hundred and six, and published in the *New Zealand Gazette* No. 41, of the thirty-first day of the same month, regulations were made, *inter alia*, prohibiting, or prescribing a close season for, for the period of five years from the date of the gazetting of the said regulations, the taking of the fish or mammal of the species known as Risso's dolphin (*Grampus griseus*) in the waters of Cook Strait or of the bays, sounds, and estuaries adjacent thereto:
> And whereas it is desirable to extend such close season during which it shall not be lawful to take the said fish or mammal in the waters hereinbefore mentioned for a further period of three years from the date of expiration of the five years prescribed in the said regulations:
> Now therefore, His Excellency the Governor of the Dominion of New Zealand in pursuance and exercise of the power and authority conferred upon him by section five of the Fisheries Act, 1908, and of all powers and authorities enabling him in that behalf, and acting by and with the advice and consent of the Executive Council of the said Dominion doth hereby make the following regulation:—
>
> REGULATION.
>
> THE close season for the fish or mammal of the species known as Risso's dolphin (*Grampus griseus*) in the waters of Cook Strait, or of the bays, sounds, and estuaries adjacent thereto, which was prescribed by the said regulations of the twenty-third day of May, one thousand nine hundred and six, is hereby extended for a period of three years from the thirty-first day of May, one thousand nine hundred and eleven, and during such close season it shall not be lawful for any person to take such fish in the said waters.
> Any person committing a breach of this regulation is liable to a fine of not less than £5 nor more than £100.
> J. F. ANDREWS,
> Clerk of the Executive Council.
>
> Printed by Whitcombe & Tombs Limited G52676

A notice by Proclamation appeared in the New Zealand Government "Gazette" of September 29, 1904, to the effect that during a period of five years from that date it would not be lawful for any person "to take the fish or mammal of the species commonly known as Risso's dolphin *(Grampus griseus)* in the waters of Cook Strait, or of the bays, sounds and estuaries adjacent there to."

A breach of this regulation was punishable with a penalty of not less than £5 nor more than £100. This proclamation (not a special Act of Parliament, as is often stated, mistakenly) was renewed, and the last "Gazette" notice on the subject, bearing date April 24, 1911, is reprinted here.

"Jack's" habit was to meet the coastal steamers trading between Wellington and Picton and Nelson, which almost daily traverse the French Pass - Te Au-miti, the "Licking (or Swirling) Current" of the Maoris. The French Pass separates the mainland of the South Island from Rangitoto or D'Urville Island, a large and rugged island rising to a height of two thousand feet.

The French Pass was so named because of its discovery and navigation in 1827 by Dumont D'Urville, the Commander of the "Astrolabe," the French corvette despatched from Toulon in the previous year on a voyage of discovery and exploration in the South Seas, and particularly on a mission to investigate the fate of La Perouse. D'Urville daringly took his ship through that narrow, dangerous, tideswept gut, quite formidable enough sometimes to full-powered steamers.

He sailed through from Croiselles Bay (On the Nelson side) into Admiralty Bay.

The "Astrolabe" struck the rocks twice and lost a portion of her false keel in this adventure. Rangitoto Island was named "D'Urville" after the gallant navigator. Pelorus Sound was named after a British brig of war, the "Pylorus," which explored a portion of it in 1838. "Jack" had for many years been identified with this Pass, but as a matter of fact he did not escort steamers through it, but frequented the northeastern or Cook Strait side of the Channel.

To see "Pelorus Jack" we used to take the steamer to Nelson. Should the steamer go direct from Wellington to Nelson, we passed, as we headed north-west across Cook Strait, the remarkable rocky islets called The Brothers, associated with Maori mythological beliefs almost as curious as those pertaining to "Kaikai-awaro."

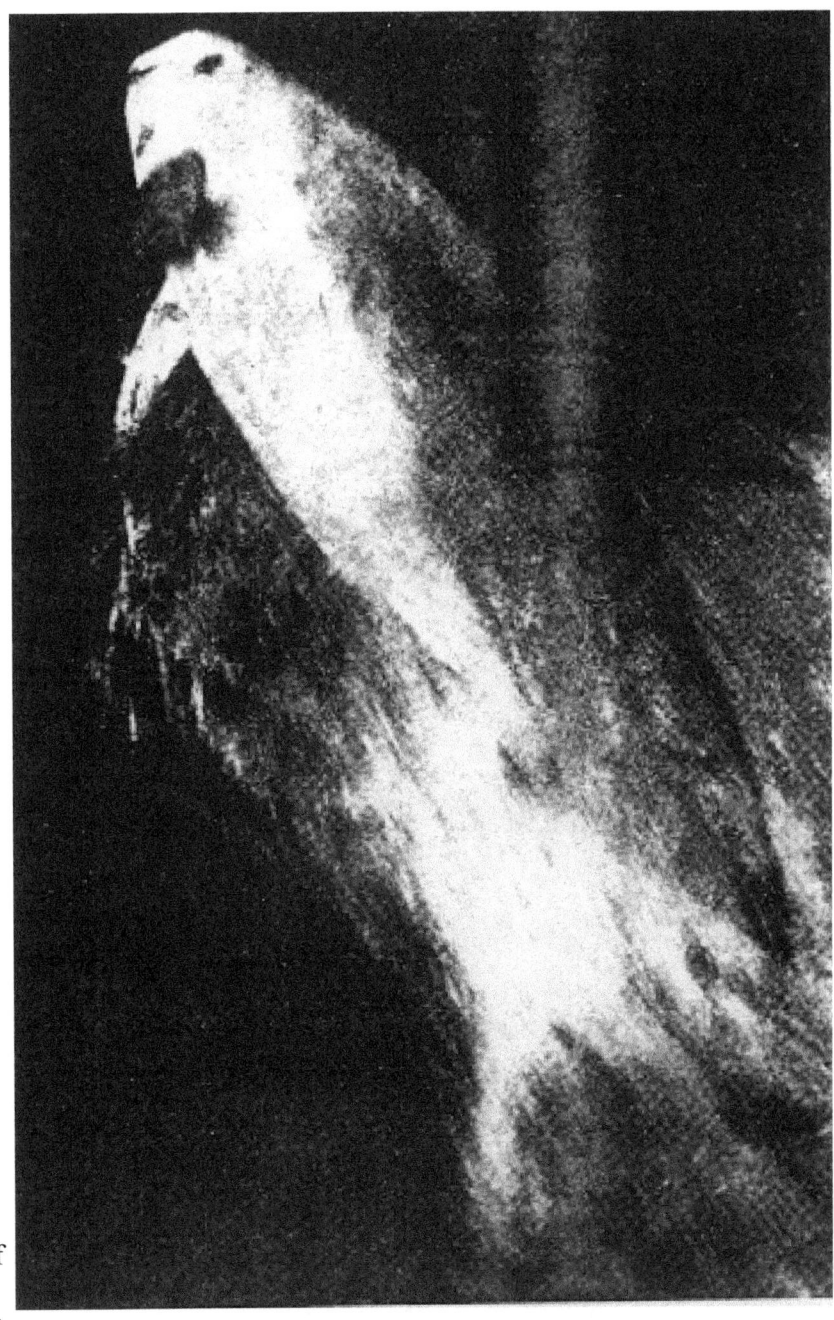

One of the few known photographs of Pelorus Jack

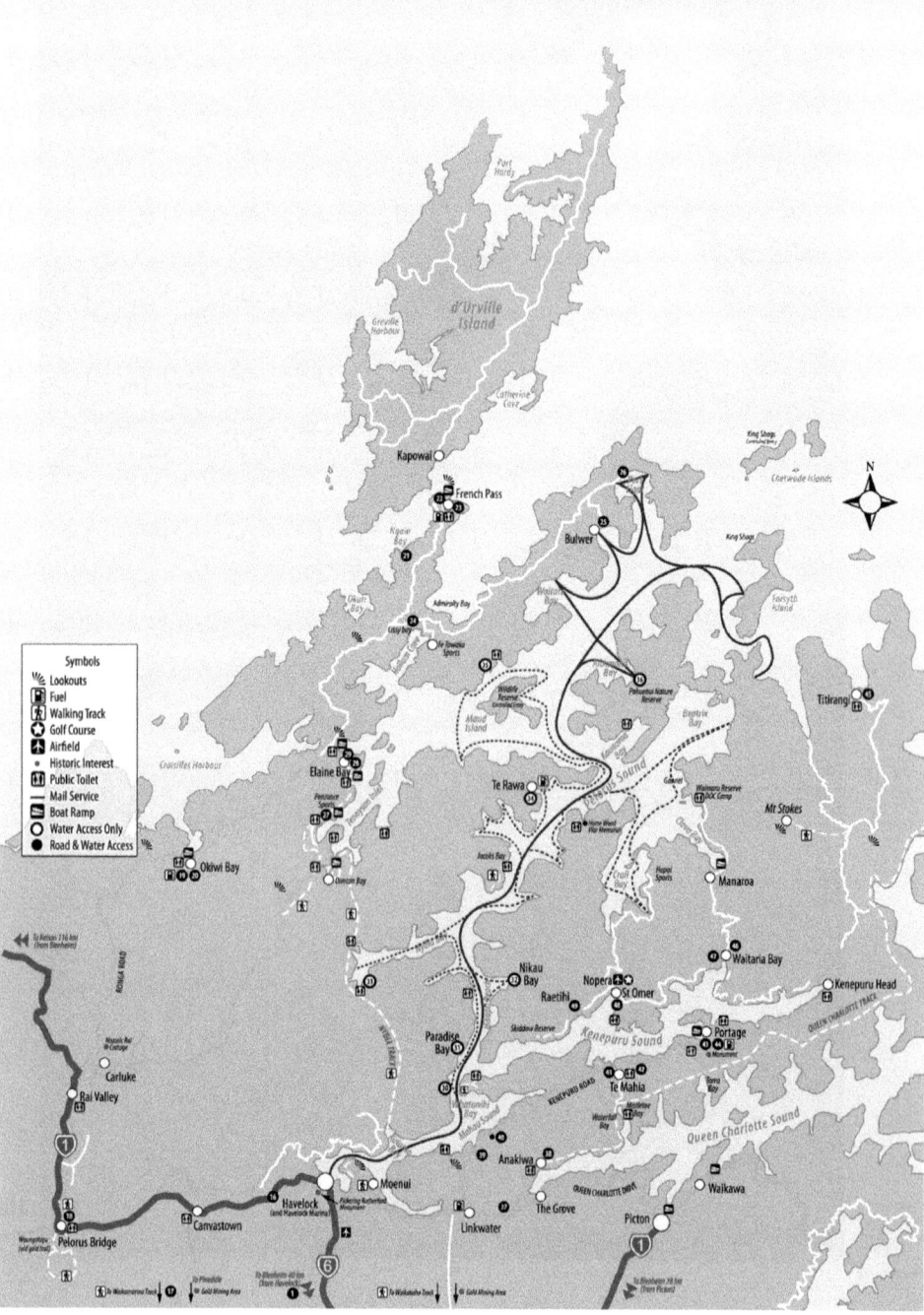

On the largest of these islets, about two and a-half miles east of Cape Koamaru (the northern point of Arapawa Island), there is a lighthouse with a ten-second flash. Nga~whatu~kai~ponu ("The Stingy flocks") the Maoris call these rocky isles with their dangerous reefs. Legend invested them with a *tapu* character centuries ago, and they were dreaded because of the perilous broken tide-rips and the high seas which were often met there. It was the custom, when strangers were crossing the Sea of Raukawa - Cook Strait - in a canoe for the first time that they should veil their eyes when approaching these rocks, and keep them covered until the *tapu* isles were past. It was customary to use three *karaka* leaves, strung together, for this purpose. An old song alludes to the eye-veiling custom - *"Kia koparetia te rerenga i Raukawa,"* in deference to the mist-wreathed spirit of the rocks. Passing the perilous isle - Captain Cook was very nearly wrecked there in the "Endeavour" in 1770 - we took a wide sweep round the broken northern coast of the country of the Sounds and presently approached the Chetwode Islets.

A small distance south of the Chetwodes, and near Alligator Head, is a small island called Motu-ngarara ("Lizard Island"), or Titi ("Mutton-bird"); off this "Jack" generally came in sight. "Here he comes!" was the cry, as the pilot fish of Pelorus-mouth came darting through the 'waves eager to meet his old steamship friend, showing his long sharp curved dorsal fin, and sometimes leaping like a porpoise right out of the water, as if in sheer joy of the diversion. He made straight for the steamer's bows, and round about and under the bows he played, sometimes flashing ahead so swiftly that the eye could scarcely follow him. The passengers crowded on the forecastle-head, and those who had cameras took more or less successful shots at the fish – snaps which generally resulted in fine pictures of salt water but no ''Jack.'' For several miles the white dolphin accompanied the steamer, as she headed towards the French Pass, swimming along with her and dashing about her bows as if he were loath to leave her. But he never actually entered the Pass. Off Clay Point, or Ana-toto ("Cave of Blood"), a conspicuous headland on the western side of the entrance to Pelorus Sound, he parted from his big friend and darted away, homeward bound.

Risso`s Dolphin

The Maoris said that he lived in a cave at Kaimahi Point, just at the entrance to the Sound, and that he seldom went many miles away from home, unless pressed for food supplies - which consisted chiefly of the cuttle-fish or octopus (the *wheke* of the Maoris). A theory advanced in explanation of "Jack's" fondness for the company of steamships was that he fed upon the octopi and squid frequently caught on the vessels' bows. Darkness did not keep "Jack" at home. He seemed to be continually on the watch. Steamers from Picton to Nelson passing the mouth of Pelorus Sound in the night as well as the day were usually joined by the big white fish, and many a passenger who left orders to be called whenever "Jack" appeared was roused out of his bunk in the midnight hours to see the finny pilot of Te Hoierc swimming quietly but swiftly along in a shimmer of phosphorescent white or plunging ahead like one of the sea-creatures in the "Rime of the Ancient Mariner," whose every track "was a flash of golden fire." The Maoris - as will be seen from Kipa Hemi 's legends - ascribed marvellous longevity to "Kaikai-a-waro." These cetaceans live to a great age, very often considerably over a century, according to scientists, but the Maori story gave "Jack" something like three centuries. However, this takes us back into the regions of mythology. Eurdpean records of "Jack" do not go quite so far.. In fact the very earliest account we have of him is one given in the Nelson "Mail" in 1911 by a veteran sea-captain. This ancient mariner, Captain W. K. Turner, of Nelson, in the beginning of the "'Seventies" was the master of a cutter, the "Southern Cross," trading to the Marlborough Sounds. One day the cutter, when on a trip to Manaroa, a bay in Pelorus Sound, was joined by a big white fish which came up and lay alongside the vessel. This was off Harding Point, at the north-west entrance to the Sound. The big creature was at first mistaken for a whale calf, and as the crew had some harpoons, it was proposed to capture the visitor. However, there were some women passengers in the cutter and they persuaded Captain Turner not to harpoon the harmless creature; so its life was spared. For twenty-four hours or more the fish remained close to the cutter.

The Lighthouse at the entrance
to French Pass c.1912

The next time he was seen was some months later; this was off Lewis Island. Here the fish remained for about twenty years. Then he moved to the mouth of the Sound and began to pay attention to passing steamers, and became well-known as "Pelorus Jack."

Captain Turner was convinced that "Pelorus Jack"' was the very fish which struck up an acquaintance with his cutter forty years before.

The only doubtful point in the story is the discrepancy between the past and present estimates of the size of 'Pelorus Jack."

Captain Turner said that the dolphin was from 25 to 30 feet in length; whereas the usual estimate of the length of "Jack" as he sported round the steamers was fourteen or fifteen feet.

THE LEGENDS of KAIKAI-A-WARO

Kipa Hemi Whird

KIPA HEMI'S *TANIWHA*.

The story of "Pelorus Jack," as the Ngati-Kuia have it, was told me by Kipa Hemi Whiro. I give here a translation of the narrative, as it came from his lips. It is hard to tell where fact ends and fiction begins; but it is the Maori story, descending through generations to the last wise man of the Sounds, who firmly believed that "Jack" was the incarnation of his ancestors' sea-god.

Kipa Hemi Whird was a white-moustached man, with a strong intelligent face and a quiet earnest manner. In 1907, when he told me these legends, he was nearly seventy years of age. He

Near the northern entrance to French Pass

had an unfortunate deformity; he was club-footed and could walk only with difficulty. (Here, it is worth noting, similarly crippled men, and also hunchbacks, were frequently set apart amongst the Maoris as tribal historians and priests.) He was regarded by all the Sounds Maoris as the leading authority on their ancestral history and *whakapapa* (genealogical recitals, and traditions). He lived with his good old mother, the ancient *kuia* Haromi (Salome) at the little native village of Okoha, at the head of Anakoha Bay, close to the mouth of Pelorus - or Hoicre, as the Maoris call this long and winding sound. His tribe was Ngati-Kuia.

The *tohunga* spoke:

"To you only of the pakeha race have I thus far told the story of my taniwha fish 'Kaikai-a-waro' [food of the deep]. It may seem strange, yet it is the story according to the Maori. My taniwha is no ordinary fish.

He is a chief in the ocean. He is the embodiment of our tribal mana; he is our family guardian God; and he has a mana-tapu, a very great mana-tapu. He is many generations old. In ancient times every Maori chief had his particular family atua or God - some were birds, some were great fish, such as whales; some were stones. *

"A long time ago - eleven generations ago [about 277 years] - my ancestor Matua-hautere first came to Te Hoiere.

He was a descendant of Kupe, the great navigator who crossed the Moana-nui-a-Kiwa (the Pacific Ocean) in his canoe Matahourua, aud who was the first, as far as our history goes, to explore these shores of Cook Strait. When Matua-hautere crossed to the South Island and the islands of the Sounds, with him, escorting his canoe, came my God-fish, 'Kaikai-a-waro.' This is my genealogy from that chief:

Matua-haittere
Matua-kuha.
Tukawae
Kuia
Wainui
Koangaurnu
Maihi
Puhipuhi
Hemi Whiro
Kipa Hemi Whiro (Born about 1840)

---

* A Ngati-Kahungunu legend says that Tuhirangi was the name of the *taniwha* fish; but Ngati-Kuia are the *tangata-whenua,* the ancient residents, and they are the best authorities on local tradition.

Perilous position of *"The Endevour"*, Captain Cook`s ship in 1770, at The Brother`s Rock, Cook Strait

Cook Strait today

"That was the first time the _taniwha_ was seen in the waters of Hoicre. The canoe crew paddled along, and 'Kaikai-a-waro' came plunging away sometimes ahead and sometimes abreast of the canoe bows, and in his rolling and diving he seemed to give the time for the paddles" - and here old Kipa Hemi imitated with his hand the rhythmic action of regular paddle strokes - "So, and so, and so - that was how he came, leading on his _rangatira_ in his great canoe. And the crew chanted their canoe songs, for they were cheered by the sight of their friendly _taniwha_. So they crossed the strait, and entered the Sound of Hoiere, and there 'Kaikai-a-waro' performed some wonderful feats.

"Swimming along ahead of Matua-hautere's canoe, 'Kaikai-a-waro' led the way up the winding Sound of Hoiere, expecting to find a clear sea-passage through between the hills. He went on until he arrived at the very head of the sound, near where Haveloek town now stands. There he nosed out the lower part of the channel in which the Pelorus River now flows; the winding course of that river near its mouth was caused by the struggle of my _taniwha_ in his attempt to force a passage through to the sea on the other side. Not far from where the sawmills now stand he turned around and made his way back to the Sound, and then he swam up the way he had come. One of the spots he stayed at while on his way to the mouth of the Sound was at a little island in the bay, which the _pakehas_ call Tennyson Inlet, in the Tawhitinui Reach of the Sound. There is a rock there, between the island and the eastern mainland, off Tawa Bay, which is _tapu_, and on the mainland opposite it is a _rua-taniwha_, or the cave of a dragon or sea-deity. In ancient times Maoris were very cautious when they went out fishing near that rock, which was a good place for catching _moki_; should they do so without observing the necessary ceremonies the _taniwha_ living in the cave opposite would become very angry, and in his wrath would capsize the fishermen's canoes.*

"Now 'Kaikai-a-waro' returned to the mouth of Hoiere Sound, and when his chief Matna-hautere settled on the shore there, not far from where I live, the _taniwha_-fish took up his abode in a sea-cave at the base of the rocky islet Kaimahi. From here he was wont to be summoned when required by means of _karakia_ pronounced by Matua-hautere and his descendants, right down to the time when Europeans first came to these parts. In the ancient days 'Kaikai-a-waro' was revered by my people, and when they went out to fish for _hapuku_ and rock-cod and their sea-god appeared they were always careful to throw him some of the fish as an offering and as food for their _taniwha_ and pilot. And 'Kaikai-a-waro' was a guide to conduct towards their destination canoes passing between Hoiere Mouth and Te Aumiti

---

* Another Ngati-Kuis legend gives a curious account of the explorations of "Kaikai-a-wsro" in search of one Thtehoto, who had followed the Pelorus River right up to its junction with a stream near the celebrated Maungstapu "Murderers' Rock." Here Tutehoto tied uli his canoe to a tree. and the stream and locality have ever aince been Icuown as Te Herenga, "the tying-up." Immense holes in three places in the Pelorus River mark the resting places of the taiiiwha, and it has always been a matter of speculation amongst the settlers why these holes never get filled with gravel, or with spoil from the banks when the floods exact their toll. Indeed, one hole, that at Brooklyn Bay, right opposite Havelock, on the western side of the Sound. and at the river's mouth, is regarded as uncanny; it is surrounded by huge gravel deposits, and at flood times these are alleged to sweep right round the pit, and but very few stones fall into it.

(French Pass) and Whakatu (where the town of Nelson now stands). He used to precede the canoes, leading them safely along the way they were to go; and he played around their bows just as he does to-day around the steamers and the launches of the pakeha.

"Now there was one of my ancestors whose name was Koangaumu. That descendant of Matua-hautere lived five or six generations ago [125 to 160 years]. He was fifth in descent from Matua-hautere. To his assistance once came his sacred taniwha-fisb. Koangaumu, when fighting against the Ngati-Tumatakokiri people, was taken prisoner by them on Nukuwaiata (the largest of the Chetwode Islands) and was shut up in a hut by his enemies, and was either to be killed or enslaved. One or two of his relatives shared his captivity. But in the night time they escaped, and hastily built a little mokihi or raft of dry flax stalks (korari) and, making rough paddles, set out upon the sea. In his dire need, Koangaumu repeated prayers to his Gods, and called upon his taniwha 'Kaikai-a-waro' for aid. And the God-fish heard, and swimming across from his sea-cave at Kaimahi he found the distressed chief and his friends float mg on their frail little raft. So 'Kaikai-a-waro' took them in charge. He swam slowly along in front of the mokihi, flashing like fire in the darkness of the sea, and joyfully the fugitives paddled along after their pilot. He led them safely southwards into smooth water and they reached the shore of the mainland on the western side of Hikurangi (Forsyth Bay). That was one of the good deeds of my ancestral taniwha."

HOW HINEPOUPOU SWAM COOK STRAIT

But more wonderful still was the story old Kipa told me of how his God-fish "Kaikai-a-waro" came to the aid of Hinepoupou, a Maori chieftainess of over a century ago, who had been marooned on Kapiti Island, and who swam across Cook Strait. It is a marvellous blending of fact and romance. No doubt it has a basis of fact, for the Maoris were remarkable swimmers, and many well-authenticated swimming feats of the coast-dwelling natives far outdo the much-sung swim of Hinemoa's across Lake Rotorua. Hinemoa 's swim from Owhata to Mokoia Island was but a matter of two miles or so.

A Maori woman once swam from Kapiti Island to the mainland, some six or seven miles, with her little daughter on her shoulders-an infinitely more difficult task than the Hinemoa exploit. But if we are to believe the story of Hinepoupon that muscular lady swam something like thirty or forty miles, across a dangerous Strait. Probably Hinepoupou crossed the Strait from Kapiti to Arapawa in a small canoe.

But Kipa and all the other Maoris insisted on the magic-aided swim, so the swim let it be. Hincpoupou was a cousin of Koangaumu, the chief of whom Kipa Hemi Whiro has already spoken, his ancestor of a hundred and twenty-five years ago. She was married to a chief named Manini-pounamu, and their home was on Rangitoto (D 'Urville Island). After a time the husband wearied of her and set eyes uj~on a new wife, and so he plotted to rid himself of his faithful Hine.

He arranged a canoe expedition to Kapiti Island, fifty miles across the Sea of Raukawa, resolving to leave his *wahine* there. He and a crew of his *hap't* set off, paddle and sail, taking

Hincpoupou with them. Arriving at the island, they camped for a space at Wharekohu Bay, which is at the southern end of Kapiti.

The perfidious husband induced his wife to walk up into the hills out of sight of the canoe landing-place, by telling her of the fragrant *kopara* plant which grew there, much prized by the Maori housewives of those times because of its sweet perfume when strewn on the matted floors of the sleeping *ivkares.*

The unsuspecting Hincpoupou set off, taking her two dogs with her, and busied herself gathering the scented *kopuri.* After the space of some hours she returned to the camping-place, but to her dismay found that her husband and his followers had disappeared. Far away at sea there was a black speck upon the blue waters-the canoe of Manini-pounamu. Poor Hinepoupou was deserted.

When Manini-pounamu returned to his home he was met by Hine 's mother, who asked him, "Where Is Hinepoupou?"

"Oh," said the faithless husband, "she met her death at Kapiti. She was killed there by the people of that island, and we had to fly for our lives."

Meanwhile, Hinepoupou sorrowed sore on the shore of Kapiti, and pondered how she should cross the dreaded Sea of Raukawa to her friends. She had no canoe, and she feared to venture to the northern end of the island, where she might be in danger from an alien tribe.

At last she resolved in her desperation to swim the Strait to Arapawa Island, the nearest point of the southern coast, where she might obtain a canoe to return to her home. It was a swim far beyond mere human endurance, but was she not a *Ariki-taniwha,* a chieftainess of sea-monsters, and could she not summon the sea-gods to her aid~

First of all she resorted to divination after the ancient manner of the Maoris in order to discover what her fate would be. Going to a flax bush she carefully plucked up the *rito-harakeke,* the heart-stalk at the root of the plant. If it broke short off it would have been a bad omen. But it did not break off short; it came out whole, and in a manner which indicated that her path lay safe and straight before her.

So with confidence in her Gods the brave chieftainess took to the sea. She walked down to the extreme southern point of Kapiti Island, and, gazing over the sea to the mountains of her island-home, she recited a *karakia* chant, an invocation to her Gods, for aid to cross the Strait, and for strength to sustain her in the ocean swim. This incantation begins:

"Ko wai, ko wai koa tera tan gata e te~e te moana?"

("Who is that yonder, wandering on the face of the ocean?") It was a very powerful incantation indeed, for it could call spirits from the vasty deep.

And having chanted her song, she threw off her flaxen mats and east herself into the sea, and stoutlicartedly struck out for Arapawa.

Hinepoupou swam on, heading south-west across the Strait. And as she swam she was buoyed up by great thoughts and potent *karakia,* or sacred incantations and charms. She recited appeals to the Maori Gods and to the sea-deities of her people, the antipodean Oceanides and Tritons.

And now the *taniwha* fish "Kaikai-a-waro" came to her aid. Far away over the sea he heard the cry of Hinepoupou to the Gods, and leaving the sheltered waters of Hoicre, he darted through the ocean, and out in the Sea of Raukawa he found the chieftainess. He guided and supported her to land, bearing her up as safely and serenely as Oberon's "mermaid on a dolphin's back." And s6 she came with her ocean-god across the swelling breast of blue Raukawa; and the seas were calmed for her by virtue of the *maro* or charm which she recited, an invocation as potent as oil upon troubled waters.

So swimming, she passed the grim rocks of Ngawhatu (now The Brothers) and presently, as she neared the mountainous coast of the South Island, she rested for a while on a half-tide rock. Here to-day you may see a round smooth rock over which, as the surf rolls in, long tendrils of sea-kelp swirl and stream and float out on the waves like a woman's hair.

The rock, in the fanciful saying of the Maori, is Hinepoupou's *llahunga* (head), and the swaying seaweed is her hair. Some Maori accounts say that Hinepoupou, piloted and uph~d by her friendly *taniwha,* first reached the land at Arapawa Island; others say that she swam to Rangitoto, her old home. Whichever place she reached, it was "Kaikai-a-waro" that brought her safe to shore. It was this very fish, "Pelorus Jack," said old Kilja.

Would you doubt the story, *pakeka* scoffer? Then go to Kapiti Island and behold the proof of its *bona fides.* There, on the cliff-side at the south end of Kapiti you will see at this day the two rocks called "Nga- Kuri-a-Hinepoupou.' They are the two dogs of Hinepoupou turned to stone. When she threw herself into the water to swim across the Sea of Raukawa her dogs were afraid to follow her and remained howling on the shore; and so they were turned to stone by the Gods.

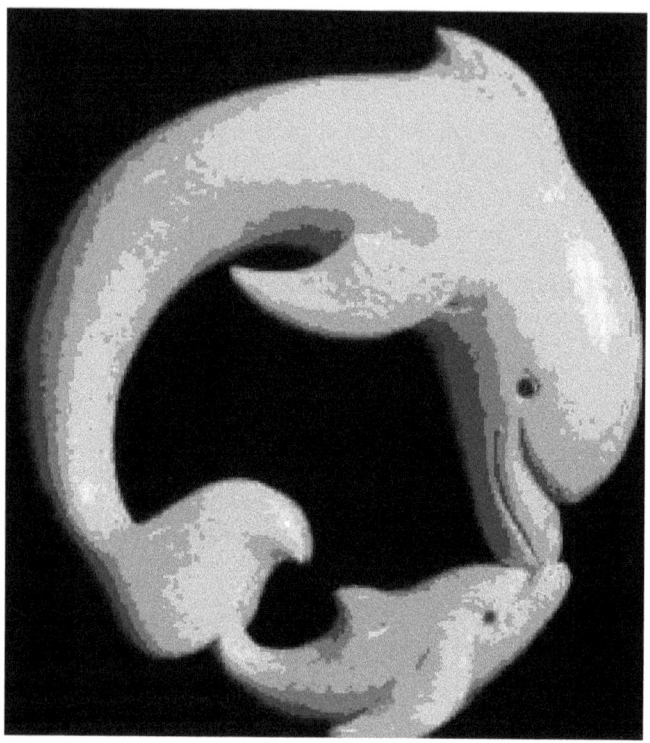

In Maori myth and tradition whales and dolphins are immortalised, often as Spirit messengers, guides, and kaitiaki or guardians. From a creation myth, Takaro and Tuteraki whanoa are said to still take whale form and keep watch on their work in re-shaping the South Island of New Zealand. Highly stylised whale carvings and painted dolphins on rock walls show the reverence given them by ancient Maori. Pelorus Jack was the world famous white dolphin, a guardian of the Ngati Kuia tribe. This carving therefore carries not only its own beauty, but also a tradition of interaction between cetaceans, people, and spirits.

There they are to be seen now on the Kapiti cliff-side, one behind and above the other with their petrified heads outstretched as if in the act of tangi-ing for their Maori mistress.

Hinepoupou 's Revenge is the sequel to the story of the great swim. Hine in due course reached her husband's village on Rangitoto Island. She concealed herself from him, and laid her plans for utu. Next day Manini-pounamu and a number of his people went out to sea in their canoes fishing for *hapuku* (groper). This was Hine's opportunity. *Karakia* after *karakia* she repeated, chants and invocations to the Gods of ocean and sky, to the sea-ruler Tangaroa, and the wind God Tawhiri-matea, and to "Kaikai-a-waro" and other genii of the deep, appealing to them to send a mighty storm and overwhelm the treacherous one. And the curse fell. A great gale of wind swept down upon the fishers, and a terrible sea arose, and Manini-pounamu's canoes were swamped and all their crews were lost. Manini's evil recoiled upon himself. "Kaitoa!"

"Serve him right!" said Hinepoupou.

Another account of Hinepoupou's revenge is to the effect that her father arranged a large expedition in canoes to fish for *hapuku* near Itangitoto Island. He so arranged matters that the faithless husbands (according to this account she had two husbands, brothers) and their crews were to anchor and fish in a dangerous part of the sea where *aniwha*, or sea-monsters abounded, while tlic people of the Island fished on the safe *haj,oku* grounds. The husbands were Taranaki men and strangers to tIle Island. He also gave orders that on the first sign of a storm the Rangitoto men were to heave their stone anchors (which were light ones) quickly and make for the shore. As for Manilimihi (msnini) - Tounamu and his brother, their canoes were provided with heavy, clumsy stone anchors and with heavy ropes which would take a long time to haul up. The fishing canoes set out, and the old *tohua* and his daughter set to work at their incantations, calling on Tawliiri.matea, the God of the winds, to send a great storm, and the *fa,jwha* ("Pelorus Jack," according to these accounts) to raise a furious sea and to overwhelm the canoes. presently the storm arose.

The Rancitoto tribespeople, forewarned, quickly hauled up or else slipped their anchors at the first indication of a brewing gale, and hurried to the shore; but the two brothers and their crews, intent on fishing-or trying to fish, for the sea-monsters scared all the *hapsktt* away - held on till too late. Then all at once the fury of *Tawhirimatea* and the sea.gods burst on them. They set to work desperately to haul up their heavy anchors, but they had hardly got them up before their canoes were overwhelmed by the great wind and the sea. They were lost, every one of them, and so Hinepoupou had her revenge.

THE TANIWHA'S LATER HISTORY

The old mall, concluding his narrative, said: "So, *pakelta,* from that day to this 'Kaikai-a-waro' has been the guardian *taniwha* of our people, the Ngati-Kuia, when they go out upon the sea. Although that fish directed or piIoted the canoes, sometilnes the children and the women when they saw him were afraid, and cried, because it was said that he was a sea mouster or *taniwkai* who swallowed human beings *(he taniwha horomi tan gata).* Not so! He was a good fish. When the white man's religion was brought to this country, the fish disappeared for a time. The Maori *tohungas* used to repeat *karakia* to him, but the incantations of the white priests frightened him away; their *karakia* were superior to those of the Maori priests. Then the Maori priests *karakia'd* that he might return and be a pilot as before for the canoes of the Maori chiefs. And in the course of after years, when fishing was carried on in these waters, this fish reappeared. When the European and Maori fishermen cleaned their fish, the remains were thrown into the sea, and were carried by the tide to the place where Kaikai-a-waro dwelt. Then he came forth attracted by the remains of the fish floating on the waters, and it was then that he began to follow the boats of the fishermen. Then after this there came the steamers, and the crews of those steamers threw meat over to feed him. This is why he first came out in search of passing steamers, a habit which he has continued up to the present time, when he is seen by us all on the steamers passing between Wellington and Nelson. "Friend, I know of the reports of these days concerning the doings of this pilot fish. I have seen him in this Sound, Hoicre; and should any white man be ignorant of the fact that this fish was seen when the mouth of Pelorus Sound was reached let me tell him that he was seen there long, long ago by my forefathers. It was my tribe, Ngati-Kuia, that owned this fish, down from my ancestors to mc and my people, and I am now his *rangatira*. He does not go through the French Pass. He generally travels seven or eight miles east and west; Kaimahi Rock is about half-way between his two]imits-Titi Island (off Alligator Head) and Nukuwaiata, Kakaho, and Para-hamuti (the Chetwode Islands) on the eastern side, and between Anatoto (Clay Point) and the eastern entrance to the French Pass. He can hear the coming of the steamers while yet far off-for is he not a God? So, friend, you see now why we revere our ancient God-fish 'Kaikai-a-waro.' And should any man doubt my story-why, I can take him out in my boat and show him the very cave where my *taniwha* live~it is only four miles from here-and perchance 'Kaikai-a-waro' himself may appear."

JACK'S DISAPPEARANCE

When Lord Bryce passed through Nelson in June, 1912, he made inquiries about this famous fish, which he hoped to see on the trip from Nelson to Wellington. He was told that the fish bad disappeared in the preceding April, and had not been seen since that time. "Jack" never reappeared. Whether he died a natural death, or was killed was never known. All that is cer-

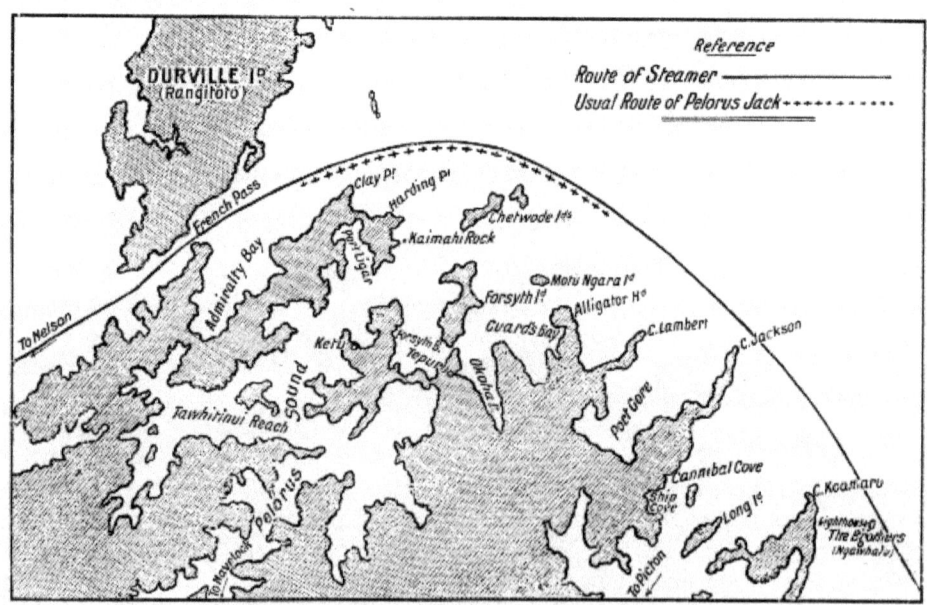

Pelorus Sound and vicinity, showing route of Pelorus Jack.

tain is that in April, 1912, he was meeting steamers near the French Pass as usual, and that he suddenly disappeared.

TU-MOREMORE THE SHARK

According to the Hawke's Bay Maoris there is a kind of Pelorus Jack, in shark form, which haunts the neighbourhood of Cape Kidnappers. This creature, a blue shark apparently about twice the length of a man, is said to have been in the Hawke's Bay waters for a very long time, and there is some local folklore about him. He has been given a name, Tu-moremore, which may be translated, very freely, as "Old Bald-head." He sometimes visits fishing parties, and if they are wise they will not begrudge him a hapuku or a rock cod as tribute. The Maoris of Pelorus paid similar respect to "Jack," but, unlike old Tu-moremore he quite deserved a fish or two for the pleasure he gave. The big shark's usual heat is from the entrance to the inner harbour at Napier down to Cape Kidnappers and back.

"Is he a man-eater?" I asked a Maori. "Oh, no," said he; the old fellow doesn't trouble the bathers. There is just one thing he dislikes, and that is the fishermen's nets, and if he finds a net in his waters he does it all the damage he can." A trawl net to Old Baldhead is like a red dress to a bull.

The Beast Of Blue Bell Hill

by

Neil Arnold

Whilst scientific boffins continue to argue over the legend of the Beast of Bodmin, big cat-flaps elsewhere are sustaining a reputation somewhat akin to the West Country phenomenon. Whilst the agricultural ministers gradually try to dissipate the image of the Exmoor beast there is something more than a moggy prowling outside my doorstep. There is a presence growing stronger in the South East whilst in the West the 'experts' are trying to disperse an enigma and as they waste their time the mystery is becoming more potent in other areas. In previous articles I have covered the Kent cats but not highlighted a certain area. However, now I have reason to do so for very recently a spate of sightings have constructed a legend that is high in the premier league of out-of-place big cats.

Judging by the press reports it seems that haunted Blue Bell Hill has been swamped by more intrigue and so with camera in hand I set off with the immortal words of Brian Glover, in *An American Werewolf In London*: 'keep to the roads, stay off the moors:'

There is something very peculiar about all these sightings of big cats. It seems that through all the private collections, zoos and local circus' there cannot be that many escapees to cause the flaps that are circulating at the moment. This is narrowed down even more by the fact that a majority of zoos are aware of what creatures are awol. Yet there are so many reports. Immediately I dismiss spectral creatures for although I am aware and believe in the black dogs etc, I just do not feel that the large felines are ghostly. At the moment they seem to be Britain's answer to Bigfoot, for although very real they are extremely elusive, hard to track and rarely leave any trace of themselves in the degree of remains or droppings. The sightings still continue though. So, the scepticism towards this furry enigma must be abolished because these beasts are in our backyards now, they are making the news and they are a frightening reality.

Blue Bell Hill, as many fortean-types will already know, is extremely haunted, as well as being a magical place. Only recently it has been sliced by a ferocious dual carriageway yet away from the rumble lies two halves of a tranquil village. The Upper Bell area, where a majority of the cat sightings have been, boasts a number of phantoms. Wizened road horrors and vanishing hikers are the most popular wraiths and witches are said to gather around Little Kits Coty. It lies at the Lower Bell and relates to the Upper Bell's Kits Coty House. They are both strange burial related structures that are completely out of place against the backdrop of the rolling fields. The former area is a collapsed monument consisting of the 'countless stones' although they are easy to count, there being twenty large, flat stones in all. These relate to other countless stones littered throughout the Medway Valley. Meanwhile, Kits Coty House looks like a miniature Stonehenge with its two upright stones stabled by a large, flat stone lying across the top in a goal-post fashion. Blue Bell Hill picnic area sits adjacent to the Upper Bell Inn. The moorland-type fields stem from a quiet country road and are only separated from the lanes by muddy ditches and boggy paths. The thickets are thorny and brambles are clustered in tight shrubland and yet the area, even in wintertime is busy with motorists who pull into the carpark at lunch times to have a snooze. It is also said that people involved in sexual affairs have been known to canoodle whilst over-looking the valley'.

My mini-expedition took me to the Hill during this year's blustery January, hoping to catch the beast. It is an area of thin grassland, ankle-breaking divots and foliage perfect for any large

DOMESTIC CAT PRINT (ACTUAL SIZE)

CAT PRINT I FOUND AT BLUEBELL HILL (ACTUAL SIZE.)

Pictures by Neil Arnold

cat. The trees are few and far between on the actual hill but gradually thicken as you reach the shrouded Burham Valley below. Here, the land is steep and inaccessible. The wind was hostile, hurtling across the landscape and biting at the face as you stand on the verge. You can see for miles into 'the glens and gorges which are the Downs. The place is hardly an equivalent to a Cornish moor but it is very scenic and sufficient enough to provide for any creature. It is not extremely vast yet it becomes more difficult to trek when entering the reservoir of Burham. The woodland stretches forever beyond this point so the cats are probably prowling for miles for fodder. On this particular day though I wanted to focus on the very recent sightings that had occurred in the actual picnic area. Upon reaching my destination I found some interesting paw-prints on a muddy path. Many people walk their dogs in this vicinity so it was easy to compare the tracks. The dog prints were very different to those I'd discovered. You could see the claws of the dogs but the other prints were very cat-like and much larger than a domestic cat. No other animal in the countryside could produce these prints, except for a mystery feline.

It was this exact spot, where only days earlier a couple from near-by Lordswood had a jumpy encounter with a big cat. Steve and Amanda Arnold (no relation to me) were walking with their two toddler children when a black beast sprang from the undergrowth.

Steve said, "It was pure black. It didn't look at us, we just heard a rustling and as I looked to my right I saw this thing pouncing downward, going very fast. It was over in seconds"

The creature was described as being as big as an Alsatian. Amanda added, "It was long-legged and skinny, that's what I'd say. Its ears were sort of floppy. . . short and floppy. " When asked if she'd seen anything like it before she gasped, "Never, because I screamed my head off."

The Author at the location where Steve and Amanda Arnold saw a big cat

Quentin Rose, zoologist, has been researching the cases for years and has theories. **

He said, "I've spoken to hundreds and hundreds of people and done hundreds of post-mortems on sheep and deer carcasses and I've identified twenty-seven parts of the country where there are regular, reliable sightings for leopards, Thirty-two for puma, ten for lynx, six for jungle cat, leopard cat 'and ocelot and four for wolverine."

When asked about how the South East features in this picture he replied, "Quite heavily. I've got leopard and puma sightings in Kent, East Sussex , West Sussex and Surrey."

These claims back up a tale concerning Battle in Sussex where a puma has been seen on several occasions, most recently by a visitor to Battle Abbey. The man saw the creature in the grounds of the gift shop and told staff of his sighting. Custodian Darryl Burchmore said, "The witness was adamant he'd seen the beast only one-hundred yards away. "

Quentin Rose believes it is up to the government to take action now because there have been six attacks on people in the U.K. He warned, "Animals, particularly leopards, that are wounded, maimed and unable to hunt their normal prey are renowned in their countries of origin such as Africa and Asia, as going for people because we are easy to catch and good to eat."

With Kent being such a rural place there obviously has to be food for the predators. However, small farms dotted in the area are not heavily stocked on sheep so maybe horses are another option. No attacks have been reported though. On my journey alone though I came across many rampant chickens, bouncing bunnies, domestic cats and birds, some of these being quite unusual. Maybe the cats will end up hunting humans but in the winter months most folk stay in their cars so the cats would have to come out of their territory.

Where such a large number of cats could be hiding adds to the mystery. Although the wooded areas are reasonably sheltered the trees are often spindly, especially during winter, so dens, marshland and tunnels must be in the reckoning. However, despite being a windswept and desolate area, many of the fields lead to 'built up areas that are part of villages. An out of the way cafe over-looks the Lower Bell and people even sit on the hill puma- spotting.

Despite being a small village the trudge from Upper to Lower is quite draining and I was lost in the labyrinth of wooded paths, steep declines, winding steps and eerie lanes but my enthusiasm was fired by a dropping which I came across. It was certainly not of a dog and too big to belong to a domestic cats bowels. The excrement was hard and crumbled to reveal what looked like cherry-pips. Of course, these creatures snack on whatever takes their fancy and the

---

** Sadly, Quentin Rose, who was one of the 'good guys' in British big cat research, despite some of the nonsense that has been published about him, died in the early part of the new century. He never got his chance to clear his name of a series of quite ludicrous allegations that had been made about him.

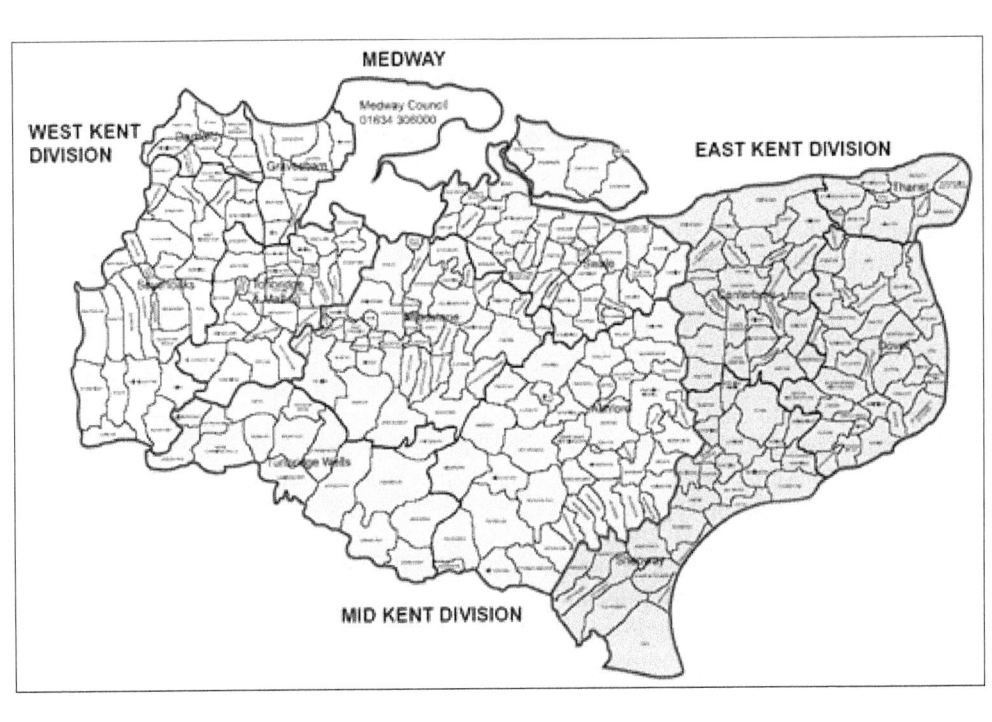

dropping didn't compare to that of a fox.

There is no telling as to when the Kent cat-flap began, especially in the areas close to my home. In the '70s a friend of mine walked down his garden path on one dark night. His own cat was by his side but soon hurried for cover when something, dark ran across his path. Whatever it was it was unseen in its speed :~ it, ". . screamed like a banshee. "Also, in the same area a number of domestic cats have gone missing. We can't put this down to a mystery predator straight away but unless stolen or run down, there is no other reason for their vanishing acts.

Around the Summer of '97 the Blue Bell Hill creature began to make the headlines. A woman in Strood reported seeing the creature drinking from a river behind her home. Through her kitchen window she saw the cat frighten cattle in a field as it loped away. Another of the better sightings involved Julie and Chris Bartel of Hempstead. They were in the area with their three children, Elysia, Susannah and Joshua. They saw the creature about five hundred yards away as it strolled across a field and then slipped into the thickets.

Mrs Bartel said, "We were walking along a hilltop and just saw this thing in the field below - it was a big, black creature. It wasn't a horse or a donkey; we couldn't make out what it was. "

However, her ten-year old son was certain it was a big cat. It had to be a big cat because of the way it was walking along

and the way its tail was hanging - it seemed to be a black panther. " Mrs Bartel went on, "There must be something out there if several people have reported seeing it. "

Local papers seem to class the reports as 'rumours' and Kent Wildlife Trust inspectors claim that the existence of the big cat is unlikely for they believe there are no significant reports of wild animals attacking farmers' livestock. What do they mean by 'no significant reports'? Does this mean they have had some kind of report but one that they can't confirm, or are they just getting themselves muddled? What else could cause this type of kill? Well, wild dogs are hardly known in the area and if they were decent enough to go by reports then they'd surely realise that the cat-flap is more evident than a wild dog-flap. The mainstream groups are certainly unhelpful in terms of this sort of thing. And their claims are weakened by a report from Christmas '97 when six sheep were mauled at the Holborough. Marshes Nature Reserve, near Snodland. The kills seemed to be of the big cat kind and locals found it very rare for livestock to have been attacked at all. Many other sheep of the flock were injured as they tumbled into a near-by river, it seemed as though something had chased them. The Kent Wildlife Trust still believe it was the work of a dog despite the fact that the 'kill' does not look like a dog mauling.

Around the same time of the Bartel family sighting, a motorist on the A29 saw a large, cat-like creature prowling on a grassy verge. The driver stopped on the hard shoulder and observed the creature as it wandered through a field.

One of the most unusual reports though concerns a man who thought he saw a gorilla in the area. Was this a bizarre misidentification of something or simply more evidence to support the existence of out-of-place animals?

During late autumn 1997, my friend, Joe Chester (who is in his sixties) was just about to settle down for sleep at 12:15 am. His cat was asleep on the back of the sofa, and they were both startled by a strange cry that had come from the valley near-by. Joe knew this was an extraordinary call and his cat just sat there, frozen. He said, "There suddenly came a painful cry, echoing across the valley. My cat knew it was something different. It was as if there was a zoo near-by and the big cats were wailing …but there is no zoo."

My venture pulled in a few local folk who seemed unaware of the big cat situation. They either don't watch the news or just don't believe it. One lady said, "It sounds like an amazing story. Are you hoping to catch sight of it? It's nice, a shame people never had a camera. I'll have to start bringing mine out with me"

Considering the picnic area is small I was surprised at how the people I met had not heard of the creatures. The ghosts of the area are known across the globe. As the day grew darker and colder a gloomy sky blanketed the greenery and I knew my hunt for today was drawing to a close. I thought I'd get a few words from the local warden who drops by to close the car-park gates.

Unfortunately, the interview was cut short by his ignorance. I approached. "Hello there, any recent sightings of these big cats?"

Immediately he scoffed, "Ha, ha, ha. No, no. I've been all over the place up here and haven't seen anything. . "

I butted in, "There was something on the news the other day. "

He replied, "Yeah, that's right . . . ha, ha!" And he stomped off.

And with that my hopes of some advice on the area blew away in the evening winds. So I took the road home as the car park emptied. With dripping nose and frozen fingers I looked forward to a hot meal. Days later I managed to get in touch with a local newspaper reporter covering the current cat-flap. He'd visited the place with his two dogs to survey the area, and although he never saw the beast he was very interested in my tales. He had been a sceptic before but as his phone never stopped ringing he had more belief. Like me, he'd returned from his venture muddy and numb but very keen. He shared some reports with me.

⁰n the 28th January 1998 a big cat had been seen by a woman in the Maidstone area. She saw the large feline from her bedroom window as it prowled the grounds of an orchard. The lady was positive she'd seen the panther.

Other reports come from the past. A few years ago, one night, a man in his car was driving up Rochester High Street and was startled to see a big cat lingering near the train station. The theory here is that the cat probably travels along the shrubbed railway lines before coming into busy areas at night and Rochester is a busy area. Known for its Charles Dickens connections, the high street links up with Strood at one end and with Chatham at the other, both known for their busy high streets. Although this sighting was at night there are still various pubs and nightclubs dotted about so this proves that the cats are willing to travel anywhere through their curiosity.

A Gillingham man said he saw a big cat emerge from the undergrowth whilst waiting for a train at Higham Station. The sighting occurred March 1997 and at the time the station was deserted. The man said, "I just stood there looking at it for about two minutes and its face was gorgeous. I have no doubt it was a puma. I did not think about it at the time but afterwards I realised that if I'd been closer it could have gone for me. "

Other sightings of the cat-creature have been reported around Capstone Country Park, Chatham. Known for its artificial ski-slope it is also a nature orientated setting with its lake and small farm. The big cat has been seen here on a couple of occasions.

With the place attracting so many people through ski-ing, fishing and general visits it could only be a matter of time before a person gets too close. This of course is not being unfair to the creature's ways but in an alien environment and such a busy area the cats may become alarmed at such activity. Their curiosity may be their undoing.

During mid-1996, Sandra Jennings of Strood was in the fields of Cobham Girls School with her children when they spotted a black panther-type creature not far away. The event alarmed the family so much that Sandra has vowed not to return to the area.

Towards the end of January the mystery feline was twice spotted at East Farley Station at Maidstone.

ONE OF THE MYSTERY CATS OF KENT, PHOTOGRAPHED NEAR ROCHESTER.

AUTHORS ENLARGED SKETCH OF THE BEAST IN THE PHOTO. THE COLOUR PHOTO ACTUALLY SHOWS A BROWNISH CAT BUT HARRY MATTHEWS BELIEVES THE CREATURE TO BE BLACK.

PHOTO BY HARRY MATTHEWS.

Other sightings have emerged from Maidstone, Susan Whitnell saw one of the cats from her bedroom window at Upper Fant Road whilst a big cat ran in front of a woman's car near the Cobham Golf Club. Also, North Dane Way, a stretch of road leading to Lordswood has also been the prowling ground for a mystery cat. This area is not far from Capstone (mentioned earlier).

Very recently though, the area of Cooling, near Rochester has been alive with big cat reports, one of these being the best pieces of evidence to date.

Harry Matthews, who works for the RSPB caught the beast of Medway on camera near the wetlands of the Thames Estuary. The shrubland is a haven for wildlife and Mr. Matthews, who lives in Herne Bay, is very is familiar with the animals of Britain. He always has his camera with him and has sighted mystery cats a few times but on this occasion, at the beginning of February he saw the cat thirty-feet away and snapped it.

Harry said: "It was very large, a black cat, certainly not a domestic oat or the sort you would want sitting on your lap"

He continued: "The first time I got close enough to take a photo of it, it was not happy to see me at all and when it heard the camera shutter it took off like a rocket. "

When asked about his theory in regards to the species, he replied, "I believe it to be an Italian wild cat, my colleague knows of such a beast, and it may be a serval but it was about the size of a fox. "

Hairy went on to describe the habitat. "We have woodland around us and a lot of open grassland where sheep graze but I wouldn't have thought it was big enough to eat a sheep. I imagine it could survive on the smaller wildlife, like voles and that sort of thing."

When asked about the cats elusive nature he said, "I think the cats just don't want to be seen, I'm sure they are avoiding humans, their natural enemy and when people do see the beasts they don't have a camera. "

However, he shows no concern at working near the creatures.

He joked: "I've asked the company for danger money but I don't think it's dangerous. "

In the colour photographs of the cat it shows a very brown creature. The first picture shows a muscular feline larger than a normal cat but Harry claims it was much blacker. The beast could well be a hybrid of species that seems the case when we look at the second photo of the oat springing away. This photo seems to show a wild cat, more domestic than out-of-place and experts at the Port Lympne Zoo, outside Hythe, don't believe that the photo shows a large big cat. They feel it could be an Abyssinian style domestic although they can't confirm this. The experts certainly do not rule out the existence of these predators though. The fact is, Mr. Matthews' photo's show a cat not associated with rural Kent.

Harry Matthews' Second picture

In my opinion the photographs are pretty convincing, especially the first, which shows the beast, near a tree. Its form in the shoulder, head and tail certainly goes beyond that of a domestic cat. The second picture could well show a different feline although the pictures were supposedly taken at the same time. This leaping cat is still of muscular frame, about the size of a fox, darkish brown and either showing faint stripes or spots on its hind legs. This of course could be a cub or a new species evolving from interbreeding.

A mystery predator has menaced the Cliffe Boarding Cattery in Cooling Street, Cliffe for a number of nights during the winter. The past three months have been unsettling times for owner Vanessa Fisher who runs the place. She is convinced that her business has become the prowling ground for the same mystery feline roaming Blue Bell Hill and the other areas. The woman has been woken by the glare of her security light and each time she claims to have caught a fleeting glimpse of a panther-type beast, the most recent sighting occurring at the beginning of February. Vanessa believes that the cat is attracted by the scent of her own cats and the food which she gives them.

Vanessa said: "Each time I have woken up in fright when the security light has come on, I have looked out the window and seen a panther which is several times larger than a normal cat. I have just sat there in amazement - too afraid to go outside."

Thankfully the cats she keeps are kept in a safe area but whatever the case, a mystery feline is still stalking the cattery.

More evidence for the big cats emerged quite soon after Harry Matthews had snapped the strange feline. A Chatham man named John Turner was walking his dog in the Blue Bell Hill area near the Robin Hood Public House when he stumbled upon a mangled fox carcass. John was stunned and sickened by the grisly find. I telephoned him in regards to his experience, which occurred February 9th '98.

He said "There is a footpath that runs next to the pub and I walk my dog there about four times a week. This time I saw the dead fox, it had been torn apart by a ferocious animal, its head and front half of its body had been ripped off by a powerful killer and it was very fresh. You could see the blood still wet and sticky, I didn't like to look at it too long because it was gory, all the intestines were hanging out, you could see the liver and everything. I just cannot think what sort of animal would do this and then I remembered the panther stories but I don't really believe in it. Another fox couldn't have done it or any wild dog and I don't think a person could have done it or hit it with a car and thrown it because there would have been pieces everywhere. I've been back since and the corpse has been moved a few inches, maybe rats have been at it or whatever killed it had returned. I just don't know what to believe, I've never seen anything like it and I've walked my dog there for a very long time. "

After my conversation with the fifty-six year-old man I decided to go back to the area with my friend Marc Ruddy. On the 12th February, a fresh and misty day, we tried to locate the dismembered fox but found no trace of it. The carcass could have been anywhere along the dark pathway. We decided to down a few pints in the Robin Hood Hub but the barmaid was quick to dismiss the existence of the cats.

She said: "I just don't believe in it at all, the photograph just shows a normal cat. "

Our extensive search throughout the day found nothing, although we managed to explore thick areas we never thought we could get to. The steep quarries of the Downs were penetrable in some areas although very dangerous. There are about five of these huge crater-like areas, constructed of chalk and running wild with pheasants and squirrels. At one point we were frozen with fear at the sound of a heavy rustling but we put it down to a scuffling squirrel and just hoped a large feline was not going to pounce in an area where there is nowhere to run.

The day was made frustrating by the fact that my trusty camera, complete with new batteries, failed to function but the magic of the Downs proved to be a quest on its own. This sort of place is deceptive, one moment you are next to a roaring motorway and then you are a world away, in a time-slip as you over-look the chalk pits and stare into the woodland world so tranquil. The only thing that separates you from a vast descent is a rickety barbed fence. There was an unsettling feeling in the air as we walked dirt tracks and grass verges just waiting for a glimpse of that black coat. As fate has it though, someone not even interested in the big cats will probably see it before me even though I'm the only person truly tracking these beasts on a daily basis. However, this eerie dwelling is frightening at twilight and what chance is there of seeing a big cat when you can't even see your hand in front of your face?

Despite local wardens not seeing the big cats I believe their existence can no longer be questioned. Recent reports of Essex beasts mauling geese have hit the headlines nation-wide and so I now plan to spend a night or two shacked up in a tent in order to at least hear these cats. Until then, the reports mount along with my excitement and the mystery thickens on a daily basis seems that Kent has a Cryptozoological puzzle with staying power whereas in the past a lot of monster mysteries have made brief appearances and never returned. The walking fir-cone of Ramsgate, the Hythe Mothman and in 1993 the closest thing we've had to Nessie made an appearance in the Thames Estuary. Several witnesses saw a long-necked creature swimming towards the Essex area but it never reared its head again. This time though it seems as if we have a mystery creature that is in no rush to move on. That is fine, just as long as the cats don't mind sharing their territory with an obsessive crypto-colleague.

"That cat's so sly, slick and subtle. . ."
Thin Lizzy

With thanks to:

Geoff Maynard,
Joe Chester,
Terry Cameron,
Marc Ruddy,
the witnesses
& my girls. xxx

ADDENDUM

As this article was completed I expected the big cat reports to continue; at times the strong perseverance of the mystery has in fact caused me to rewrite parts. I believe that the Kent cat-flap will indeed cause me to write further articles, but instead of waiting until next time I would like to add two further cases pertaining to cat kills which came to my attention upon completion of this piece. On February 12th a sheep was found dead in a farmers field at Wrotham Heath, near West Malling. The carcass was found by a woman walking her dog. The sheep had a large chunk of flesh torn from its back.

She said: "I didn't like to look at it because it was a horrible sight. The injuries were too savage to have been done ~ a dog - it must have been a large cat. "

Meanwhile, on Tuesday, February 17th at Lydd Golf Club, head green-keeper Simon Grand was shocked to find a massacre among swans. Two birds were completely disembowelled and the other, still alive, almost had its wing ripped from its body. The damage was too severe to have been done by foxes, and the surviving bird had to have its wing removed after it was rescued by a Surrey swan sanctuary.

Rescue co-ordinator Chris Evans said: "When we got to the scene it was like someone had cut a duvet of feathers and shaken it - It was a disturbing sight. It was obviously something powerful with speed that killed them. Swans themselves are powerful birds."

Sanctuary owner Dorothy Beeson said: "It was not a domestic dog or foxes, I have never seen anything like this before, this was nothing less than a massacre."

Mr. Grand has worked at the Golf Course more than four years and was appalled at the carnage.

"In the main kill area there are large paw prints, they are three to four inches long and quite deep. It had a long stride."He added: "The swans are a big part of the golf course and to find them like that was upsetting. "

To be continued. . . without a doubt.

*****

Waitoreke:
The Enigma from New Zealand

By Craig Heinselman

New Zealand, the Island nation in the South Pacific Ocean, separated from the super-continent of Gondwanaland (Gondwana) and all other landmasses for eighty (80) million years has evolved a unique biosphere. With only bats as the indigenous land mammals the avian fauna predominated, yet reports of a mammal living in the mountain lakes and rivers has been reported over the years. What then is this animal, the waitoreke?

Known to the natives in various incarnations as kaureke and waitoreke (various spellings waitoreke, waitoreki, and waitoteke), yet with varying descriptions from otter-like, beaver-like and seal-like in habitat and characterization, have occurred from South Island, New Zealand for over two hundred years. The more descriptions offered the more enigmatic the reality of this animal becomes. Also creating en enigma is how could this animal, this mammal (one agreement is that it is mammalian in characteristic) have arrived on an island isolated for millions of years, yet with no fossil record in existence?

That is the chore ahead. To evaluate the various theories as to what the waitoreke is, and through that process determine the most likely identification as to what the waitoreke is.

We shall do so by looking at the etymology of the name of this animal, the diversities of life on and around New Zealand, the habitat of the animal, the theories (through behaviour, anatomy and habitat) and the evidence thus far provided by the witnesses and chroniclers of the waitoreke.

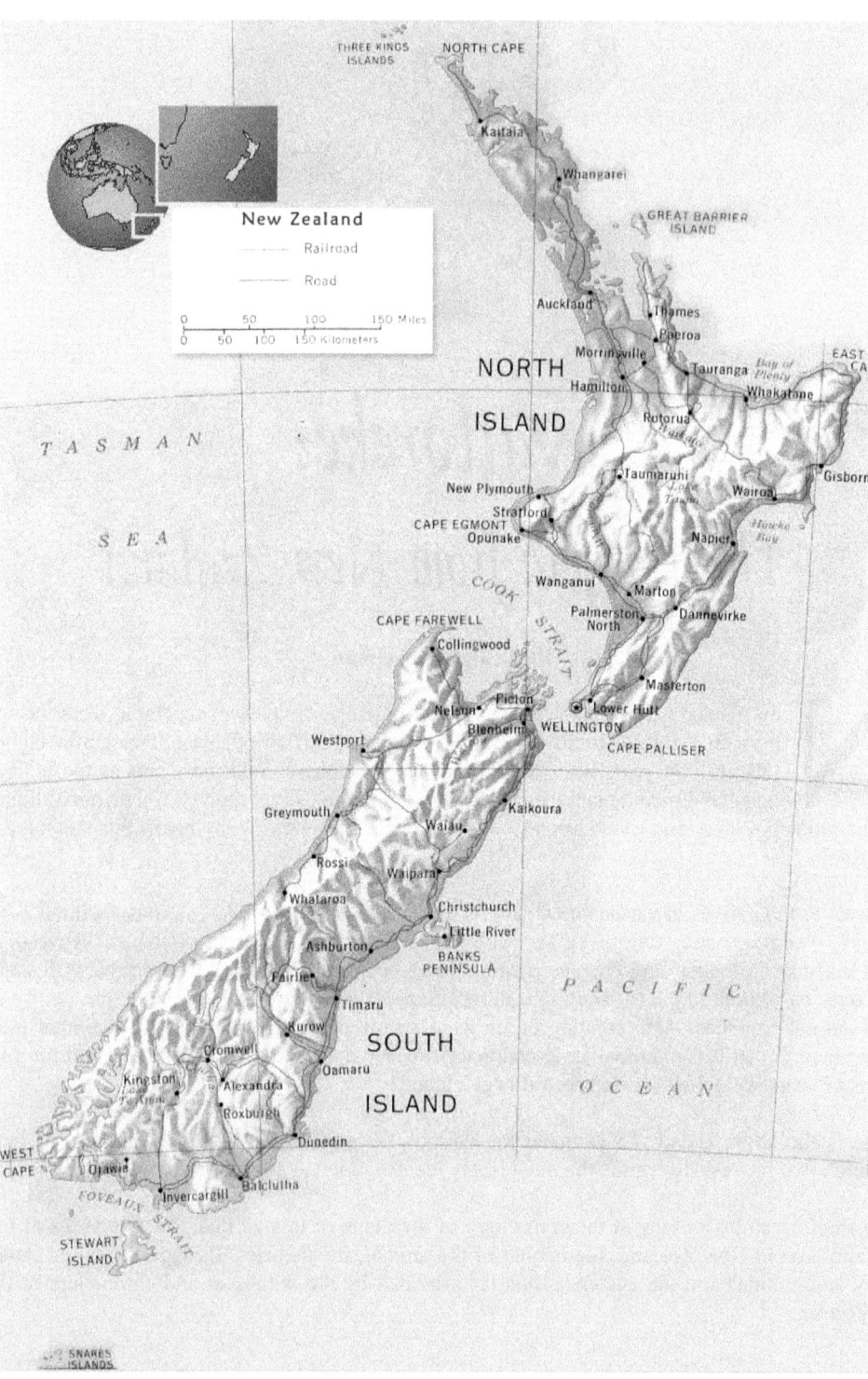

Biological Diversity on New Zealand

New Zealand formed from the super-continent of Gondwanaland. Gondwanaland is the Southern Hemispheres super-continent made up of what are know known as the continents of Africa, Antarctica, Australia and South America as well as India, Madagascar and New Zealand. Gondwanaland and its counterpart Laurasia (for the Northern Hemisphere) once joined as part of Pangea, the mega-continent. Through tectonic plate shifts the landmasses drifted apart (as they do still today) Roughly 80 million years ago New Zealand separated from Gondwanaland along the edge of what would become Australia. Australia separated about 30 million years latter from Antarctica.

Since the time frame separating New Zealand was greater than of its neighbor Australia the faunas where not the same. The animal species that migrated to the islands did so be wing, in such a time as the distances between New Zealand and other coastal areas was close to each other. New Zealand boasts some of the oldest terrestrial life on Earth. The beech forests originated in South America, and the ones on New Zealand (as once part of South America via the super-continent) are perhaps the longest surviving forests on Earth.

Additionally New Zealand boasts the tuatara (*Sphenodon punctatus* and *Spehodon guntheri*) an archaic reptile that has virtually remained unaltered for over 200 million years. The primitive genus Leiopelma frogs (*Leipelma hochstetteri, L. hamiltoni* and *L. archeyi*) are found in New Zealand as well. These constitute the oldest lineage of frogs alive today, sharing characteristics with fish, and lacking characteristics associated with other frogs (aside from an American cousin *Ascaphus truei*).

Flightless birds like the moa, kiwi, and kakapo have evolved in this landscape. What were lacking were indigenous mammalian carnivores. There were several bat species present on the mainland: the New Zealand short tailed bats (*Mystacina tuberculata* and *Mystacina robusta*) and the lobe lipped bat of New Zealand (*Chalinolobus tuberculatus*) as well as later the arrival of pinnipeds like the New Zealand sea lion (*Phocartos hooker*) and the New Zealand fur seal (*Arctocephalus forsteri*). There may also have been a few small rodents native to the island (as fossil records of a shrew like mammal 115 million years old was found in 1997 in Australia). But lacking were the larger carnivores associated with other continental landmasses.

It was not until humans began to arrive on the islands that the animal diversities began to grow. With the arrival of the Maori (various dates on their arrival abound, typically 1000 - 2000 years ago) the ecosystem changed. The Polynesian Rat, or Kiore (*Rattus exulans)* was introduced, as were dogs.

Later, when explorers like James Cook and European colonists arrived in the 1700s, livestock and foreign animals where brought. Nowadays there are many mammals present in New Zealand, amongst them seven species of deer, wild hogs, wallabies, possums, and quolls - to name a few. These new inhabitants now threaten the original wildlife, and in some cases have either wiped them out, or brought them to the brink of extinction.

Etymology of the Waitoreke

What does waitoreke mean in a language aside from the Maori language (the natives of New Zealand)? John Colarusso offers that the word waitoreke and kaureke that ended with "reke" are the closest of the various spellings for translation. "reke" meaning quill or spur (knob and bone), and when put with other Maori words waitoreke translates (roughly) 'out to water diver (with the) spurs'. Kaureke would translate to 'many spurs', with extension of kaurehe meaning 'monster' and perhaps' tuatara'.

For all intents and purposes, the term waitoreke will be used here out for this cryptid animal. Although kaureke may be a valid name (or a separate animal altogether) as well, the popular and more common waitoreke will be suitable.

The Habitat of the Waitoreke

The habitat of the waitoreke varies little from report to report. The animal is associated with water, and is seen in it or just beside it. What *does* vary are the reports of what the waitoreke inhabits; is it a lodge like a beaver? Or a tunnel system like some otters, and the platypus?

In 1855 Reverend Richard Taylor's book *Te Ika A Maui, or, New Zealand and Its Inhabitants*, was published. In it is the following note:

"A man named Seymour, or Otaki, stated that he had repeatedly seen an animal in the Middle Island [Note: Middle Island is actually modern South Island], near Dusky Bay, on the southwest coast, which he called a musk-rat, from the strong smell it emitted. He said, its tail was thick, and resembled the ripe pirori, the fruit of the kie-kie, which is not unlike in appearance to the tail of a beaver. This account was corroborated by Tamihana te Rauparaha, who spoke of it as being more than double the size of the Norway rat, and as having a large flat tail. A man named Tom Crib, who had been engaged in whaling and sealing in the neighborhood of Dusky Bay for more than twenty-five years, said he had not himself seen the beaver, but had several times met with the habitations, and had been surprised by seeing little streams dammed up, and houses like bee-hives erected on one side, having two entrances, one from above and the other below the dam. One of the Camerons, who lived at Kaiwarawara, when the settlers first came to Wellington, stated that he saw one of these large rats and pursued it, but it took to the water, and dived out of sight."

In this account we have references to a beaver type lodge. Yet, in 1921 one A.E. Trapper witnessed an animal while on a bridge crossing the Waikiwi River. Shortly after he found a hole in a bank in the location the animal disappeared in. And again in 1973 a G. Pollock, who had been researching the animal, found a tunnel system in the reeds of a swamp on the Taieri Plain. These two shelters described match different known animals.

So the question of living habitat is troublesome. The one item that can be agreed upon is that aside from waterways, the waitoreke lives in higher elevations towards the southern portion of South Island. Elevations vary from sea level to 3764 meters (12,349 feet) at Mount Cook in the Southern Alps across the island, with lakes varying in altitudes. There are some exceptions to this statement, as caption James Cook's crew described seeing an animal along the coast in Dusky Sound: in 1773:

"A four-footed animal was seen by three or four of our people; but is no two gave the same description of it, I cannot say what kind it is. All, however, agreed that it was about the size of a cat, with short legs, and of a mouse-colour. One of the seamen, and he who had the best view of it, said it had a bushy tail, and was most like a jackal of any animal he knew."

The Usual Suspects: The Otter Theory

River Otter

The theory that an otter is responsible for the reports of the waitoreke is the most popular and common. In the 1867 Ferdinand van Hochstetter writes in his book on New Zealand:

" My friend Haast writes me about the Waitoreki under the date of the 6th of June 1861 as follows: "3500 feet above sea level I saw at the upper Ashburton River, in an areawhere no

human foot ever walked before me, its tracks on many occasions. The tracks resemble those of our European otter but are somewhat smaller. The animal itself was seen by two gentlemen who own a sheep ranch at the shore of Lake Heron in the neighborhood of the Ashburton River at an elevation of 2100 feet above sea level. They describe the animal of being of a dark brown colour, of the same size as a large rabbit. They hit it with a whip. It emitted a whistling sound and disappeared quickly in the water among the weeds."

Additional reports also detail some anatomical and behavioural characteristics.

In 1957 a woman saw an animal near the Aparima River that was described as having small pop eyes and flat round ears. The neck was hidden, had fur like a cat and short whiskers on its face. In 1971 a hunter, familiar with NZ wildlife, watched an animal slide down a bank of the Hollyford River for a period of about fifteen minutes. This animal was described as smooth, short brown fur, small head with no visible neck or ears, tapering thick tail, and 91-107 cm (3 - 3.5 ft.).

Another witness in the early 1970's saw the animal eating a fish, the webbing on its feet was visible. In 1971 tracks the size of matchboxes, with indications of webbing, where found in a swamp on the Taieri Plain (same area that Pollock later found the tunnel system).

Otters constitute a taxonomic status having thirteen known species, however none is known south of the Wallace Line. The most likely candidate based on the description is of a river otter. As this otter, unlike the sea otter (*Enhydra lutris*), is a freshwater variety. Additionally the river otters do from time to time come in close proximity to oceanic environments, especially along the shorelines (as the Chillian variety, *Lutra felina,* often demonstrates). The river otters closely match the waitoreke. The have a brownish coat of short dense fur, a rounded head, short necks, thick tapering tail, short legs, webbed feet and small ears. Their size varies from 76 - 132 cm (2.5 to 4.3 ft) including the tail. River otters rarely travel beyond a few hundred meters of a water body and live in burrows in close proximity to the water. Additionally otters have been known to travel many miles overland to find rivers, and in doing so they travel by running and sliding. Additionally Walter Mantell records in 1838 an interview with Tarawhatta of the Ngatimamoes:

"He informed me that the length of the animal is about two feet from the point of the nose to the root of the tail; the fur grisly brown, thick short legs, bushy tail, head between that of a dog and a cat, lives in holes, the food of the land kind is lizards, of the amphibious kind, fish-does not lay eggs."

Again this closely matches the description of the waitoreke. Where as the main substance of the otter is fish, however they are also known to eat small reptiles, birds and mammals. This would be particularly important if the otter were to travel over land from one river to another.

The other possibility in the otter theory is that a member of the clawless otter genus *aonyx* may be responsible for the waitoreke descriptions. These otters typically range in size from 60 - 171 cm and are similar in appearance to the river otter. However, one noticeable difference is that clawless otters lack the clear cut webbing on their feet and posses smaller claws.

This is an important consideration as the waitoreke has been described as having webbing on its feet (sighting in 1970's near Opihi River, and track finds in the same time frame on the Taieri Plains).

If the otter is to be considered then it had to have traveled across the ocean. This could occur in one of two ways, either it was brought to New Zealand or it swam there on its own accord. G. Pollock offers the theory that Indonesians visited New Zealand before the arrival of the Europeans. As river otters are often trained to catch fish, they were on board with the Indonesians and either escaped or where released. Thus, a male and a female at a minimum had to have escaped in order to create a viable population. The other possibility is that otters became caught in a current and either swam or floated on flotsam over the distances from another continent to Oceana.

The Usual Suspects: The Pinniped Theory

New Zealand Fur Seal:

The pinnipeds are those marine mammals the seals, sea lions and walruses. Their distributions are world wide, and of all the theories as to what the waitoreke is the pinniped have a leg up ahead of time. These mammals are present in New Zealand on their own. There are three pinnipeds that stand out as possibilities the New Zealand sea lion (*Phocartos hookeri*), the New Zealand fur seal (*Arctocephalus forsteri*), and the southern elephant seal (*Mirounga leonina*)

The New Zealand sea lion, also called Hooker's sea lion, grow from 160 - 250 cm (5.2 - 8.2 ft.)depending on the sex. The males are the only ones that exhibit a brownish coat, as the females show a grayish one. They have short muzzles and round heads. Their feet are not made for terrestrial migrations or long travels on the ground, as they have adapted flippers. These sea lions are mostly segregated to coastal areas, however they have been known to travel inland a couple miles during breeding season. They are also of the family *otariidae* which are categorized as having visible ears, occasional freshwater habitats and pronounced sexual dimorphism.

The New Zealand fur seal grows from 130 - 250 cm (4.2 - 8.2 ft) depending on the sex. Their necks are large and they exhibit a brownish fur. Again their legs are made of flippers, making land excursions short. They too are of the family *otariidae* and as such show similar characteristics as the New Zealand sea lion. Predominately the food source for the fur seal is of oceanic nature, being squid, octopus and fish.

The southern elephant seal is much larger than the sea lion or fur seal, reaching lengths from 200-600 cm (6.5 - 19 ft.). As such there size alone greatly reduces their chance of being the waitoreke.

There are some other pinnipeds that do reach New Zealand on occasion, these being the crabeater seal (*Lobodon carcinophagus*) with a length of 203-262 cm (6.6-8.6 ft.). They have a slim body with a long muzzle. Its main food source being krill, and living predominately in Antarctica. They belong to the family of *phocidae* and as such show characteristics such as no external ears, some freshwater habitats and variable sexual dimorphism. The leopard seal (*Hydrurga leptonyx*) also occasionally arrives in New Zealand from its Antarctic home. Characterized by being 300-380 cm (9.8-12.5 ft) long and closely resembling the crabeater seal. They are also of the family *phocidae*. The weddell seal (*Leptonychotes weddelli*) also occasionally leaves Antarctica and arrives in New Zealand. Their sizes are comparable to the leopard seal.

Although adapted in most cases for a freshwater habitat, pinnipeds almost exclusively live in the oceans. There are some that live landlocked in Russia, as well as some that will travel upstream into freshwater lakes (as in Loch Ness). Although adaptable, their known sizes in that region of the world is much larger than that described to the waitoreke. Additionally the burrows (not lodges) ascribed to the waitoreke are not behaviourally consistent with the pinnipeds. The footprints found, with webbing also indicate a terrestrial locomotion, of which pinnipeds are limited due to flippers instead of feet or webbed feet. In 1948 H. von Haast printed the following report by Sir Julius von Haast (who previously supplied Ferdinand van Hochstetter's 1861 report) in *The Life and Times of Sir Julius von Haast:*

"Traces of a quadruped of smaller size, of nocturnal habits, and the stride which was between seven and eight inches, and indicates that its mode of progress was by jumps or springs, was discovered by me in the riverbed of the Hopkins, the stream which forms Lake Ohau, and as there is every reason to believe that this animal still exists in great numbers, hundreds of tracks having been found in one night in the fresh-fallen snow, we may hope that some specimens of this entirely unknown quadruped will soon be obtained.".

The Usual Suspects: The Monotreme Theory

Echidna

Monotremes are the egg laying mammals, encompassing the platypus (*Ornithorynchidae anatinus*) and the echidna (*Tachyglossus aculeatus* and *Zaglossus bruijni*). Having archaic reptilian characteristics, such as shelled eggs, skeletal structure and excretory system (end of the intestines, genial ducts and excretory ducts share one single chamber), the monotremes are considered the most primitive form of mammals alive now.

Identification of the waitoreke through the monotreme line is difficult. First the echidna needs to be discarded, as its physical appearance is alien to the descriptions of the waitoreke, having an anteater like snout and porcupine like quills. The platypus is also posses a problem identifi-

cation, as its appearance is that of a biological jigsaw puzzle. Having the tail of a beaver and the snout of a duck. Such an animal stands out in ones memory, if only because of its difference. Two characteristics do match that of the platypus; its feet are webbed as described by witnesses, and it does have the right fur to match the description.

Yet, there are some reports of the waitoreke stating it lays eggs as Te Taumutu states to Walter Mantell in 1838. But, that is the end of the correlation. There is fossil evidence of monotremes dating back 100 million years ago as demonstrated by the fossilized jaw and teeth of *Steropodon galmani* from New South Wales, Australia (1985) and the jawbone of *Kollikodon* ritchie (1995). This time frame of the discovered fossils fits the time frame of the separation of New Zealand from Gondwanaland, and offers the possibility that New Zealand once fostered monotremata forms of life. If this lineage evolved, then the possibility arises that a fourth species of monotremes exists and is responsible for the waitoreke sightings.

Platypus. From François Péron's *Voyage de découvertes aux Terres Australes.*

Physical Evidence of the Waitoreke

Does there exist any physical evidence for the waitoreke? Sir Julies von Haast reportedly obtained a skin of the waitoreke in 1868. It was in poor shape, but is described as brown with white spots lacking webbing between the toes. Unfortunately this does not offer definitive proof for the existence of the waitoreke. In all likelihood the skin was of a variety of quoll, which where released in New Zealand in 1868. The quoll are carnivorous marsupials from

Australia of which all known species have a brown coat and distinguishing white spots on their skin

Track finds are the next and final physical evidence left. Although circumstantial, they do offer some important clues. Tracks found in the Taieri Plain swamp are described as showing webbing and being matchbox size. Sir Julias von Haast had stated that that the stride of the waitoreke was seven to eight inches. Of all the animals theorized as the possible cause of the waitoreke, the track finds described point toward that of the otter or beaver.

River otters typically show a slight webbing in their tracks of the hind feet, but seldom is highly visible. An adult otter normal has a foot spread of 3 1/4 - 4 inches in width and length, with a varying stride depending on the terrain and movement of 10 to 15 inches. Compared to a pinniped track which offers no discernable foot, but a shuffle of earth. The beaver offers a distinct webbing pattern in prints and is hind feet are 2 1/2 - 3 inches wide by 5 -6 inches long.

So the identification of the waitoreke through tracks is just as troublesome as through observation.

What then is the Waitoreke?

What is the waitoreke? Even after looking at the various theories it is a troublesome question to answer. Characteristics match those of various known animals, but not all characteristics match a specific animal. Of all the animals evaluated, the otter is the best fit. The behaviours described such as sliding and diving are characteristic of the otter. The physical descriptions closely match the otter (more so than the other animals theorized). The habitat also closely matches that for the otter, with tunnel systems for living and the ability to travel long distances over land.

The native people describe two animals, one amphibious the other land dwelling. This matches the otter closely. As an animal traveling far from water (cross-country) can be associated as a different animal than one observed in the water. The smell of the musk reported on several occasions also matches the otter. Vocalizations (such as described by Ferdinand van Hochstetter) also match the otter, which is capable of a wide range of guttural sounds.

We also know that early mammals where present prior to New Zealand separating from Gondwanaland. These include some early monotremes and placental mammals. Through evolution a convergent species could have also arisen in New Zealand to fill in an ecological gap. Additionally some of the reports could be misidentifications of known animals. Small fur seals could be misidentified, and lead to the reports of the waitoreke originating from the coastal areas. Another possibility is that the early reports (prior to the 1800's) are those of something unknown, but the more recent ones are of an escaped animal kept captive during colonization of New Zealand by the Europeans. With such an introduction of an alien species, the waitoreke's ecosystem was altered and it was no longer able to cope and survive. The alien species thus takes over the system, and throws another twist on the tale of the waitoreke.

These are all possibilities, and none can be completely proven one way or the other. Further study of the area is needed. For if the waitoreke is ever found, its nature could change how the mammalian family tree is shaped or how the historical immigration of people to New Zealand is viewed.

Selected Sources:

Smith, Malcolm Bunyips & Bigfoots, Millenium Books (1996)
McDougall, Len The Complete Tracker, Lyon Press (1997)
Ley, Willey Exotic Zoology, Viking Press (1959)
Costello, Peter In Search of Lake monsters, Coward, McCann & Geoghegan (1974)
Heuvelmans, Bernard On the Track of Unknown Animals Third Edition, Kegan Paul International (1995)
Riedman, Marianne The Pinnipeds, University of California Press (1990)
Burton, Maurice Living Fossils, Readers Union (1956)
Shuker, Karl The Lost Ark: New & Rediscovered Animals of the 20th Century, Harper Collins (1993)
Nowak, Ronald M. Walker's Mammals of the World 5th Edition Vol. I and II, John Hopkins University Press (1991)
Wallace, Alfred Russel, Island Life, Part of Prometheus's Great Minds Series, Prometheus Books (1998)
Murie, Olaus J. Animal Tracks 2nd Edition , Part of Peterson Field Guide Series, Houghton Mifflin Company (1974)
Straham, Ronald A Photographic Guide to Mammals of Australia, Ralph Curtis Books (1995)
Australia / New Zealand Map from World Cart
Hammond New Headline World Atlas, Hammond Incorporated (1998)
Webster's concise World Atlas, Barnes & Noble Books (1997)
Barrett, Cahrles The Bunyip, Mail Newspapers Ltd. (1946)
Ritvo, Harriet The Platypus and the Mermaid and Other Figments of the Classifying Imagination, Harvard University   Press (1998)
Mayr, Ernst Evolution and the Diversity of Life (Selected Essays), Harvard University Press (1997)
Ridley, Mark (Editor) Evolution, Oxford University Press (1997)
Quaternary fossil faunas from caves in Takaka Valley and on Takaka Hill, northwest Nelson, South Island, New Zealand by T.H. Worthy and R.N. Holdaway, *Journal of the Royal Society of New Zealand Vol. 24, No. 3, Sept. 1994,*
Tooth Marks of History by Mark Hurrell, *The Times June 10, 1992*
Shaking the Family Tree by Kate Wong, *Scientific American January 26. 1998 (Online)*
Mystery of the mammal that shouldn't be here by Leigh Dayton, *Sydney Morning Herald November 22, 1997*

In the Shadow of Wolf's Castle

by Roy Kerridge

Wolves may no longer live at Wolf's Castle Crag in Pembrokeshire, Wales, but at least one man believes that the fields and woods below the hill are home to an equally fearsome animal, the Wolverine. At home in the northern pine forests and tundra of Canada, Scandinavia, Russia, Siberia and Alaska, the Wolverine is a great shaggy brown weasel, as ferocious as its tiny counterpart, though as big as a medium sized dog. Its powerful jaws, revealing still more powerful teeth in an ominous grin, recall those of a Tasmanian Devil. French Canadian and Red Indian fur trappers have reason to call the Wolverine itself a "devil", for the wily animal reaches trapped mink and marten before the hunters arrive, and devours the animals - the trappers' livelihood. Nor can the cunning Wolverine easily be caught in a trap. Stories of its trickery and defiance of Man are told in tents and wooden cabins all over the North.

David Lawton-Watts, a retired auctioneer, lives on an estate landscaped by himself from bleak fields at Letterston, within sight of Wolf's Castle. Not long ago I went to see him there. After showing me a convincing plaster cast of a large round clawed pawmark, Mr. Watts told me of his first sight of the wolverine.

"I was out in the Landrover at night, when I saw a dark shape with a pair of blazing bright eyes shining in the darkness", he said. "I couldn't think what it was! Then not many days later, I saw an extraordinary animal in the fields. I know badgers and polecats very well, but this was much bigger! When it ran, it looked exactly like a bear. I have no doubt in my mind but that it was a wolverine. Where it has escaped from, I can't imagine!

"Neighbours have seen it several times, sometimes mistaking it for a badger until they see it run and notice the sheer size of it. I once had a very good view of it at night by torch-light. I've been collecting reports of the animal and some people have seen two, one slightly smaller than the other. Possibly the second animal has been run over, as someone I know saw an unknown brown animal lying motionless by the roadside. When he went back to investigate, a few hours

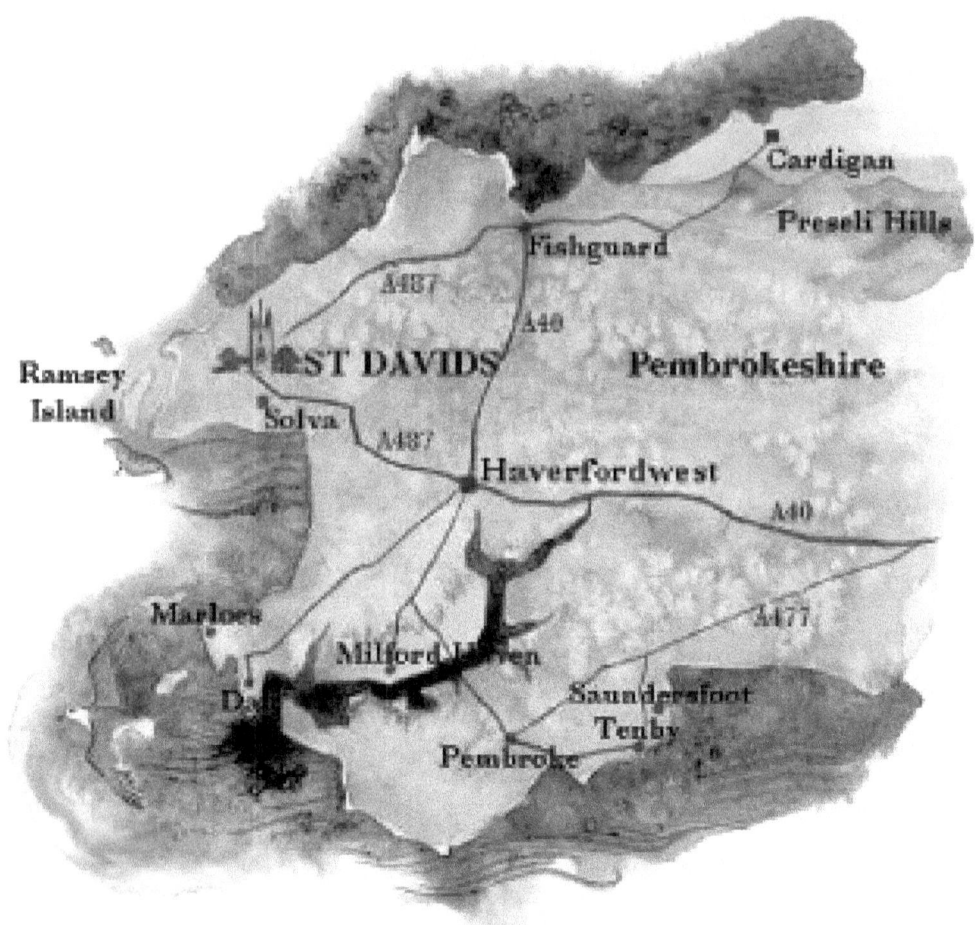

later, there was nothing there.

"I have now grown quite used to the idea of sharing my land with a wolverine, but at first I was very fearful that it might attack someone. You can just imagine - a farmer casually asks his little girl to bring a tool from the outbuilding, she goes into the barn or shed, and gets attacked by a wolverine hiding there. So I had hundreds of leaflets printed and circulated around the farms, schools and villages:

'Beware of the Wolverine.'"Now I think the alarm is over, as it's become clear that the wolverine is quite wild and shuns the habitation of Man. It feeds mainly on carrion, and seldom kills. There is a lot of fallen stock just left in the fields these days, that would in times past have been given to the Hunt. This provides food for the wolverine, yet every so often, the animal seems to get overcome by blood-lust and kills one or more lambs, tearing them apart. Don't ask me why! I have had a lake made on my property, and when I drained it, I found very clear wolverine footmarks in the mud. Before making a plaster cast, I photographed the footprints."

Mr. Watts kindly gave me a copy of a photographed pawmark beside a ruler, to show the scale. Like the wolverine itself, the print was half the size again of that of a badger. Later, I compared it with a picture of a wolverine print in *Guide to American Animals*, and it looked exactly the same.

When shown around the schools, complete with pictures, Mr. Watts's alarmist wolverine pamphlet caused a small sensation. Many children were delighted to be told that they lived in wolverine country, and did school projects on the beast. Donna Thomas, a teenage schoolgirl from nearby Sealyham Home Farm, told me that a wolverine was her "favourite animal." In fact, a wolverine, if taken as a cub, can be tamed and makes a delightful pet! I have heard of one that was trained to sit up with a cap on its head and a pipe in its mouth! In zoos, wolverines bound up and down their enclosures in a lively display of energy. Incidentally, Sealyham dogs were first bred at Donna's farm for the purpose of routing out badgers, but I doubt if one would make much impression on a wolverine!

Leading me outside to his Landrover, Mr. Watts took me on a bumpy safari over the fields to the lake where the wolverine had been seen. We rollocked over tussocky grass like a motorised wolverine. Soon we reached a fine artificial lake, long and wide, its bulldozer birth scars long grassed over. Beyond the lake, up on a rise, tall fir trees suggested a Swedish horizon.

"Those trees must make the animal feel at home," Mr. Watts commented. "I daresay it climbs them when anyone goes near."

A72YJW Alamy Images

AR81RA Alamy Images

A wolverine has a touch of pine marten, as well as of badger, in its make-up, and can climb a tree as nimbly as a small bear. Judging by the tricks it plays on campers in the Far North, stealing knives, forks, clocks and other property and burying them nearby, it has something of a bear's sense of humour.

At the far end of the lake, by a shed, we dismounted and my companion showed me the trap he had set for the wolverine. It was made on the principle of a "cage" mousetrap; the meat inside balanced to set off a trigger that would cause the door to snap shut. So far, Mr. Watts had caught foxes, badgers, cats and a neighbour's collie, all unharmed, but no wolverine. To my mind, the trap seemed too small, the door liable to hurt the wolverine's back if it shut. Would the animal enter a space which gave it no room to turn around?

"He reached in, and lifted out the meat without disturbing the trigger", Mr. Watts told me, in grudging admiration. "I scattered smaller pieces of meat outside the trap, to make it seem more natural." Foxes and badgers certainly appreciated this generosity, for I could clearly see their clawed pawprints in the mud. The fox-prints were neat and slender, the badger-marks squat and rounded, like their owners.

"What do you plan to do with it if you catch it?" I asked.

"I've already had an offer from a wildlife park at Neath, near Swansea. They've got a cage ready, so I'll sedate the animal and take it up there."

This struck me as being somewhat over-optimistic, given the frustrating experiences of hardened Hudson Bay trappers in pursuit of the wily wolverine.

"See that spot there, in the undergrowth," Lawton-Watts pointed. "That's where I first saw him clearly by daylight. The ring of pale fur on his back really stood out - it glowed golden in the sun."

It seemed clear to me, from this description, that Mr. Lawton-Watts definitely had seen a wolverine.

Over the 'phone, I spoke to another wild beast-chaser of the Wolf's Castle region, Mr. Bernard Davies, a retired Pembrokeshire National Park Ranger, who lives near Fishguard. In cloak and dagger fashion, he arranged to meet me at a seaside car park.

"I'll be in a silver courser", he explained.

"Whatever's that? I asked, my mind racing. I've heard of a Cream Coloured Courser, a rare bird akin to a plover, but not a Silver Courser. And if it's a bird, how do you wear it?

"It's a car!" he said shortly, so I at once passed the 'phone to my motoring friend Dave Arthur, the eminent banjo-player. Dave drove me to a rainswept car park, and soon identified the car (whose name might be spelled 'Corsair'). Bernard Davies proved to be a strong weather-beaten man who had worked outdoors for most of his life. His eyebrows rivalled Wolf's Castle itself in cragginess, and his tattooed hands were gnarled. In one of them he held up a formidable plaster cast with long claw prints, which we all agreed had been made from the large bear-like hind paw of a wolverine.

As if capping a party trick with a still better trick, he then produced two large plaster casts of enormous round paw prints. One, of vivid orange plaster, had been made by a naturalist friend from a paw print at Exmoor, supposedly that of the Beast of Exmoor itself. The other, a much larger pawmark, had been made locally, from a print discovered by Mr. Davies. It measured five inches by four.

"Pumas!" he announced.

We adjourned to a garage cafe', where he produced sheaves of documents and went into details. It appeared that Mrs. Davies objected to wild beast chasers tramping over her carpet and disturbing a tea party she had arranged, hence the outdoor meeting.

Bernard Davies holding plaster casts of pawprints
Cast on left: "Beast of Exmoor" made by Nigel Brierley (author of "They Stalked by Night") Cast on right: "Beast of Pembrokeshire" made by Bernard Davies

"I made this plaster cast from a footprint left by an animal that raided Penysgwarn Farm, at Croesgoch, only four miles up the road", Bernard Davies told me gravely. "That's Tom Evans's place. The puma or whatever-it-was struck in the early hours of the morning of June the twelfth 1997. It got into an open-ended barn where the calves were kept. Tom Evans heard the dog barking, but he didn't think any more of it. The animal ignored a number of sick calves that were there, and killed a fine healthy bull calf. Then it dragged the body right out of the barn, seemingly by the tail, as the tail was half pulled off when they found the remains. It dragged the calf out into the field, under an eight bar metal gate, and there devoured it."

Bernard Davies

Apparently it was a classic puma kill, a death-bite on the neck and the ribs stripped of flesh. Dogs break off bones, if they can, and chew them. Big cats rasp them clean with their scouring tongues.

Even a domestic cat has a "cheese grater" tongue, as you can feel with your finger. The calf had been sent to the local branch of MAFF, the Ministry of Agriculture, Fisheries and Food, and the usual fobbing-off or cover-up was expected. All over Britain, from Dorset to Durham, farmers are complaining of big cats attacking stock, only to have their complaints pooh-poohed by officialdom.

Big cats in Britain have been reported occasionally over the past hundred and fifty years, but sightings skyrocketed after the Dangerous Wild Animals Act of 1976. This law penalised keepers of exotic pets, and caused many such animals to be turned loose in the wild.

"There was a Cardiganshire scrap dealer who owned three tame pumas that ran about all over his land," Mr. Davies remarked. "When the Act was passed, the pumas disappeared and he couldn't account for them. He said he'd had them put down, but no one could trace the vet who did it. Now Cardiganshire is the source of most wild big cat reports around here. Come to mention it, I've heard of someone who still keeps Caracal Lynxes, Act or no Act. But those haven't escaped, luckily."

(Caracal Lynxes are scrub-desert animals of striking beauty, and would certainly feel unlucky if forced to brave the wet climate of West Wales).

"The first report of a big cat round here I knows of came from Hays Castle in 1971," Mr. Davies continued. "Reports built up in the late 'seventies. I can't remember the year exactly, but about that time, before the road below Wolf's Castle had been widened, I was driving along at quarter past eight in the morning, when a big cat, the size of a puma, crossed the road a hundred yards in front of me! It was jet black all over - a wonderful sight! I couldn't stop for long, as traffic was coming up behind me and I had to move on. It was ever since then I've took an interest in the big cats, though I've only gone into it properly since 1986."

So saying, he produced a map of West Wales marked military-style with "sightings". Round stickers made a polka-dot pattern that concentrated on Wolf's Castle and North Cardiganshire. Different colours denoted Black Big Cats, Brown Cats (pumas) and Black and Brown Cats, also places where sheep or calves had been allegedly killed by the cats.

"Reports seem to show that there are at least four Big Cats at large in Pembrokeshire - two brown ones and two black ones. Not to mention our old friend the wolverine!"

Among the many photographs of slaughtered farm animals in Mr. Davies' collection were some pictures taken recently at a Lower Tregennis farm. There a mystery animal had torn a jagged hole in a firmly nailed up hardboard window and bitten the heads off three geese kept inside. No fox or dog could have ripped open hardboard so quickly, breaking through the middle of the board.

"Only the jaws of a wolverine could have done that," I opined, and Mr. Davies agreed. Forty sheep had been killed in the countryside around nearby Solva in the space of six months, and Mr. Davies seemed inclined to blame the wolverine. I doubted this, since big cats had been seen in the district. Lawton-Watts thought the wolverine killed lambs sometimes, but not sheep. The more prosaic MAFF view, that rogue dogs are the killers, cannot always be ruled out, even in the romantic setting of Wolf's Castle!

Bernard Davies identified the brown big cats as "pumas" and the black cats as "black panthers." I dispute this, mildly enough, as I've become convinced that a damp climate, inbreeding or perhaps cross-breeding with other species has brought into being a hitherto-unknown animal, a Black Puma, or British Big Cat. A black panther is a very formidable animal, far more ferocious than a puma, and if panthers were on the loose dropping from trees onto picnickers, reports would not be of dead sheep but of dead people. Dark brown and even piebald big cats have been reported, suggesting that pumas are developing regional colour varieties.

At Whitchurch, near Solva, sheep had been found dead with their faces mauled and their ears bitten off, a characteristic of alleged "big cat kills" all over Britain. Des Townley, a retired engineer from London who died recently, took two colour photographs of the Beast of Solva. I examined these pictures, now in the possession of Mr. Davies.

They showed an animal that was undeniably a Big Cat pacing along a field below a gorsey bank. Bernard Davies had later been photographed against the same bank as a guide to relative sizes. The Solva Cat, by the Bernard standard, thus proved to be the size of the proverbial Labrador, but much more nimble and lightly built. It was a Big Cat, but what kind? "Puma" said Mr. Davies, while Des Townley had written "Definitely a lynx."

Bright chestnut in colour, with one white forepaw, the animal boasted large pointed ears and a tail halfway in length between that of a puma and the bobtail of a lynx. The shape of the head reminded me of a serval, a lynx-like cat from Africa, said to be on the loose on the Isle of Wight. Both Davies and Townley were right, I decided - the animal was a cross between a puma and a lynx! Such crossbreeds have been reported from Exmoor. Escaped big cats aspiring to matrimony have to take what partners they can get. Both pumas and the less dangerous lynxes were kept as pets before the 1976 Act. These two cats, unlike most "exotics", hail from northern climes and are well able to survive a moorland winter.

Despite his slight Welsh accent and air of a man very much at home in his surroundings, Mr. Davies, I learned, only moved to Wales in 1954. He is from Worksop and his wife from Taunton. A Welsh accent is easily acquired, and a surprising amount of thoroughly Welsh Welshmen and women have one or more parents or grandparents who are English. Generalisations about Celts and Saxons are just that - generalisations. There was no Celtic mysticism about Bernard Davies, a blunt outdoor man whose life story showed him to be well-qualified to talk about wildlife. After a short spell down a coal mine, young Bernard Davies had joined the Army at the age of sixteen and a half, and served ten years with the Royal Artillery. While still in the army, he learned rudimentary gamekeeping and reared pheasants for the officers to shoot. He then left to become a full-time gamekeeper on big estates in Pembrokeshire.

"I worked for Lord Emlyn, the son of Lord Cawdor, and after that I became a Pest Control Officer for the Min. of Ag., trapping mink, doing fox-research and rabbit control. I did that for six years, then I became a National Park Ranger for twenty two years up to retirement. So I'm familiar with most British animals. Once I had to do research on otters - lovely animals! I was watching a mother otter and her two babies near a river, when my dog jumped up, ran to the otters, picked up an otter cub in its mouth and dropped it at my feet. For some minutes, I stood looking at the little otter moving around, when all of a sudden it began to scream! The old mother otter couldn't stand this, so she ran right up the bank and tried to go for my

dog! The dog ran away. Then the mother otter rushed to her cub, picked it up in her mouth, and ran back to the river where all three otters disappeared."

One report from Bernard Davies' files had been written by a warden for the Countryside Council for Wales, John E. Davis. Apparently, on the seventeenth of September, 1996, John Davis had gone fishing at Teifi Pools, Lyn Gorlan. He looked up and thought he saw his black Labrador on the opposite bank.

He whistled for the dog, but it didn't come. So he whistled again. It was just as well that the creature didn't respond, as it wasn't the dog at all! The Labrador had gone to sleep in the rushes beside its master. No, the animal John Davis saw, now very clearly, was a black Big Cat. He made "nature notes" on the beast, which I quote.

"Body: the height of a large Labrador, but longer and more slim.

Legs: longer than a Labrador dog's.
Tail: long and thick, not pointed.
One and a half times as long as a dog's.

Condition: coat very shiny and very black. Seemed in very good condition and very agile. Sheep in a field nearby seemed distressed."

Our interview concluded, Dave Arthur and I bade Bernard Davies farewell, and headed straight for Solva to see if we could photograph the Beast.

We were out of luck, though farm cats encountered on the narrow steep-banked twisty lanes made our hearts leap once in a while. A surfeit of Big Cat stories can affect the judgement. The prevalence of farm cats showed that no Big Cats were near. All the sheep seemed relaxed.

"This scrubby countryside reminds me of East Africa," Dave said wistfully. "On a road just like this, in Kenya, I once saw a cheetah resting under a bush just like that hedge."

I looked hard, but no cheetah did I see. It was time to leave Wolf's Castle, Land of the Puma and of the elusive Wolverine.

(Since writing the above, I learn with sadness of the death of David Lawton-Watts in July, 1997).

****

Did some Mesozoic reptiles evolve from flying birds?

by Allan E. Munro

In the 19th century it was commonly assumed that the dinosaurs (and some other Mesozoic animals now considered reptiles) were not like the reptiles of today but instead like birds and mammals, warm blooded and active, and placed in seperate classes with names no longer in use, such as *Saurornia*. Gradually this view went out of fashion and these animals became considered ectothermic, scaly reptiles. Later still, the old views returned for several reasons, such as the observation that dinosaurs behaved like (and fulfilled similar ecological niches to) mammals rather than modern reptiles. Birds are known to be descended from small carnivorous dinosaurs which implies that these dinosaurs were endothermic just as bats inherited an endothermic metabolism from their ancestors. Since all other dinosaurs were descended from a small, carnivorous dinosaur this implies that the whole group was endothermic like mammals. Even so, birds were seen to be the only feathered dinosaurs, with *Archaeopteryx* as an early bird.

The birds are maniraptoran dinosaurs because of the way their wrists operated, which allows birds to fly. Sinosauropteryx, a compsognathid, is one of an early offshoot of this group of dinosaurs and is now known to have been insulated by feathers, which implies that dinosaurs such as *Tyrannosaurus, Velociraptor* and *Gallimimus* were feathered (or secondarily featherless, as tyrannosaur skin imprints show no trace of feathers). The feathers of *Sinosauropteryx* are simple compared to a bird's feather, like hairs in shape. The feathers of two dinosaurs from the same region, *Protarchaeopteryx* and *Caudipteryx*, are more like those of a bird. From the position of a brooding Oviraptor, an animal with similarities in the skull to *Caudipteryx,* the animal seems to have used feathered wings to incubate eggs.

One bird feature other than feathers was held to be the wishbone, which was supposed to have evolved for the purpose of flight and is now known from many dinosaurs other than birds, not all of them maniraptoran. such as Velociraptor are very similar to archaeopterygids and it is often suggested that they are closer to modern birds than *Archaeopteryx,* which would prove that they were secondarily flightless. Certainly *Rahonavis* is sometimes held to be a *Dromaeosaur* and was definately capable of flight, with wings with feather attachments as in birds that flap their wings powerfully and also a pelvis that shares adaptations with modern birds to help ventilate the lungs (despite being vertical rather than slanting towards the back as in birds and *Dromaeosaurids*).

Some studies place various other groups of dinosaur closer to birds than *Archaeopteryx* was, which implies that they were flightless. Certainly there seems to have been a ridge for feather attachment along the arms of *Avimimus,* which implies that this animal evolved from ancestors that were powerful fliers. *Longisquama*, from the early Triassic seems to have had both feathers and a wishbone and seems to have been a dinosaur. In this form, the arms were not used in flapping flight. Two rows of feathers seem to have been used to glide instead. The animal was definately arboreal. Certainly *Unenlagia* had arms that could be raised and lowered like the wing of a bird. This implies that *Unenlagia* was arboreal and a reversed toe as used by birds when perching was present in dinosaurs as distantly related to birds as *coelophysoids*. The claws of such animals as *Archaeopteryx* and *Deinonychus* are those of animals that climbed or had climbing ancestors. Certainly dinosaur evolution took place mostly in the trees, whilst flight developed from arboreal gliding.

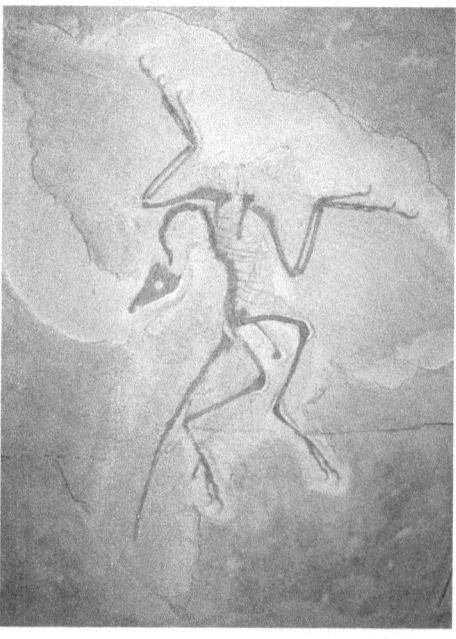

The dinosaurs seem to have inherited feathers from their ancestors. The *erythrosuchids* of the early Triassic are usually considered to be distant cousins of the dinosaurs.

They had a bone microtexture like that of dinosaurs and mammals and were also rare compared to their prey which is a sign of endothermy. They lacked certain features of dinosaurs and mammals that are associated with endothermy such as fully erect limbs and a backbone that undulated vertically more easily than horizontally and also they were not as rare compared to their prey as are carnivorous dinosaurs and mammals. The crocodilians may be more closely related to dinosaurs than were the *erythrosuchids* and indeed crocodilian anatomy shows traces that they had endothermic ancestors, such as the scutes that evolved from structures similar to feathers, and some extinct crocodilians had a more active way of life than modern forms. *Cosesaurus* is usually held to be more distantly related still to the dinosaurs than *erythrosuchids*. This animal may have had a covering similar to feathers, and may be close to the pterosaurs which also had a similar covering. Therefore, some adaptations that indicate endothermy in dinosaurs were found in quite distantly related forms. This therefore is paralel to the evolution of endothermy in mammals and their relatives.

The therapsid relatives of mammals had a similar bone microtexture and (in carnivores) scarcity compared to their prey to that of *erythrosuchid*s, and certainly lived in cold climates which implies that they were hairy. The lineage that gave rise to mammals were adapted towards small size, and so were the ancestors of dinosaurs. Both presumably were at least partially arboreal insectivores that spent some of their time in cool trees.

As they had a higher metabolism compared to their larger relatives, they developed further anatomical adaptations to an endothermic lifestyle. It is hard to say when flight developed among the dinosaurs but definately among the dinosaurs which had a reversed toe and grasping hands. At least those dinosaurs with attachments for wing feathers were definately descended from ancestors that could powerfully fly. Flight evolved from arboreal gliding to escape from predators, and if a ground-living dinosaur is not secondarily flightless then it is descended from an arboreal ancestor.

A Ghana Folk Tale

by Louis Baba
via Roy Kerridge

EDITOR'S NOTE: this story is particularly interesting to the fortean zoologist for a number of reasons.

* It has a number of similarities to the Swahili folk tales quoted by Bernard Heuvelmans in *On the track of Unknown Animals* whilst discussing the semi legendary East African mystery felid, the Mngwa or Nunda:

"*The mngwa appears in many Swahili songs and tales. Hichens quotes a war-song dating from about 1150 A.D.*

sikae muyini kewa kitu duni
Nangia mwituni haliwa na 'nngwa
(I dwell not in the city to become a worthless idler
I plunge me in the forest to be eaten by the 'mngwa!)

From the legend of the Sultan Majnun' quoted by Edward Steere in his Swahili Tales, the beast seems to perform the same function in folklore as dragons did in Europe in the Middle Ages.

One day the Sultan's cat escaped and slaughtered its master's poultry. The Sultan's soldiers asked him if they might kill the cat, but he replied: "The cat is mine, and the poultry are 'nine."

The cat went on to eat sheep and goats and, presumably growing larger and stronger on this diet, soon was eating cows, horses and camels. But each time the Sultan replied that the cat as well as its prey were his, so there was no occasion to kill the cat-until the day when the cat devoured three of the Sultan's sons. Then he changed his tune: "It is no longer a cat," he said; "its name is nunda."

LOUIS BABA

I was born at Saltpond, a coastal town in the Central Region of southern Ghana, in 1954. I am the first son, but second child of a family of three children. I started school at Gowrie in the Bongo district of the Upper Region, where I lived in the palace of my uncle, Gowrie Naba Kiribeo Zankanga.

I subsequently went through St. Charles *Primary* School to St. John's School, in Bolgatango. From Bolgatango I came down south again and got a job as departmental clerk in. the University of Cape Coast, where I did the non-degree awarding course, In-Service Training scheme. In 1977 to '78 I was at Prestea underground goldmines as a spannerman. Presently I am a game trooper for private dealers in wildlife.

The Sultan's seventh son was a wily and fearless hero. He decided to seek Out the monster and kill it. He killed a large dog and came home singing:

Man'd wee, niuldga
Nundd mia wdtu
(0 Mother, I have killed
The nunda, eater of men.)

But his mother disillusioned him, and he proceeded to kill a civet, a zebra, a giraffe, a rhinoceros and an elephant, one after the other, thinking each time that he had killed the nunda. And each time his mother told him his mistake.

At last, after announcing that he would not return without the nunda's body, he set off into the forest with his servants. And one fine day he saw the monster.

"This must be the nunda he said. "My mother has told me that its ears are small and these are small; she has told me that the nunda is broad and not long, and this is broad and not long; she has told me that it has two blotches like a civet, and this has two blotches like a civet; she has told me that its tail is thick and this tail is thick; and all the peculiarities she told me of are here."

Convinced that this time he was not mistaken, the young man shot the monster with his gun. When he came back his mother greeted him singing:

Mwanangu, ndiyzyi
Nundd mia wdtu.
(0 my son, it is he
The nunda, eater of men.) The story ends in the usual way: the youngest son inherits his father's kingdom, marries a fair wife and enjoys a long life much loved by his people.

The moral of this interesting legend might well be: Never mistake a nunda for any ordinary creature."

* Creation myths are always interesting and this story in particular explores the relationship between man and the other animals that share his environment. It many ways it is similar to various ancient irish stories cited by Karl Shuker in *The Mystery cats of the World - from Exmoor Beasts to Blue Tigers* (1989) and my own *Smaller Mystery Carnivores of the Westcountry* (1996) when discussing the putative Irish wildcat.

* This is also interesting corroboration for west African shapeshifter stories about were beasts.

* Finally, however, it is a good story, that if Louis Baba had not sent it to Roy Kerridge and then Roy Kerridge had not sent it top us otherwise, many of us would not have read.....

THE FIRST MAN.

A LEGEND OF THE UPPER EAST REGION OF GHANA.

Here is my story; and there is a man. He was the only human on earth. And always he trod the face of the earth, holding a bow in hand and a quiver of arrows over his shoulder. Honey and leaves were his food. All animals knew him and regarded him as just another creature.

Out in the field one morning, as Man roamed, dark clouds came over the sky with whiffs of chill. When lightning and thunder exploded and flashed, he considered a cave he knew. Intuitively he went there to shelter. The chill got even more severe, while the darkness deepened. He propped his torso on the wall and soon was swept by sleep.

Moments later, Hare came in, also to shelter. He cast a look at sleeping Man, deciding the directly opposite corner was most suitable not only for shelter, but also a vantage position to keep an eye on the first occupant of the cave. Next, snake came in, sow the two sleeping creatures and settled in the most dark corner of the cave. It was the Hyena. who got the last corner. When monkey came in he picked a position closer to Man.

The cave got filled at last in this sequence:

Squirrel, Finch, Gecko, Bat, to mention those who matter. No big animal came in.

Then it rained. It really poured! For two days and nights it poured without abate. Monkey hunched, Snake coiled himself, Hare hung his legs upwards. Style and pose in sleep varied from one creature to another. Bat was most impressive; he glued his legs to the roof, hung his head downwards, and still slept soundly for nights and days on...

The great downpour mellowed to a gentle drizzle on the third day, by which time everyone felt the pangs of hunger. One after another they rustled to awake. Hare elected to take change of the situation.

He asked Monkey to go out and find food. Monkey came back with the soft fruit of the baobab tree. (They are not wrong, who call this the monkey bread). This they peeled and shared.

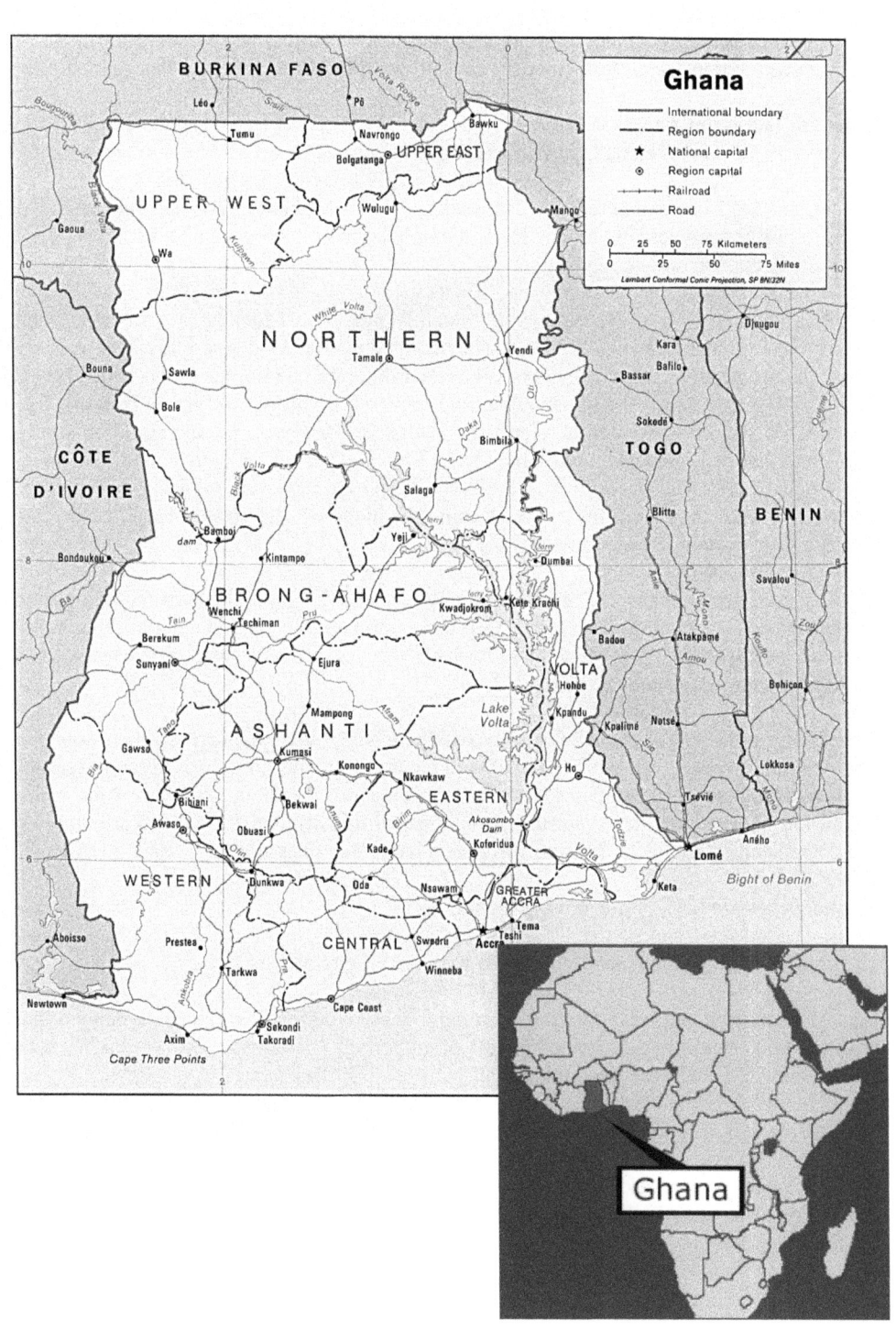

Squirrel went out and returned with nuts, which they all ate. Hare himself brought in tender leaves, which roved succulent. Hyena came back with honey, and all were pleased. Snake rolled in with herbs, which miraculously gave all warmth. (Man detected this as medicine).

When all had taken a turn, it came to Man's at last. Hare did not hesitate to put it to him. Man rose to his feet, stretched, yawned, adjusted his quiver firmly and held his bow tightly.

The instant Man had sauntered out, Hare whispered to Monkey: "You, go spy on him, and come tell us how he gets his food. He is the only creature who goes about carrying one load and holding another".

A trot, following a dash, then a swing, smoothly propelled Monkey to a treetop, where he stilled in the foliage screen, and peered with squinted eyes at Man. Presently Man sees a warthog foraging the dew-spangled shoots. Poor Warthog, he had not sheltered with Man. Carefully a right hand went over the shoulder and drew out an arrow, the arrow then got fixed in the centre of the bow, pointed at Warthog. It looked like a jerk, no one could be sure. The arrow had vanished, no one could have seen it travel. There wasn't any sound, but the Warthog toppled over. Man went over to the fallen heap. He retrieved the arrow and stabbed the point a couple of times into the earth, cleaning the blood off the arrow. He lifted the carcass over his shoulders and headed towards the cave.

Monkey had seen enough. He didn't climb down carefully. He let himself fall straight to the ground, then raced back to the cave. Back there and panting, he managed to breath out: "If you did ever see him passing a hand over his shoulder, run for your dear lives, was all he could get out, as Man now entered.

By rubbing stones, Man made a fire, over which he hung the gutted carcass, then went back to sleep out the exertion. No one uttered a word again, till the burning flesh began to reek. Hare motioned Monkey to rustle man to awake to check about the food. This time Man woke up, he felt the nuisance of flies swarming on blood that had dripped from the Warthog's wound and coagulated on his back. Naturally he sent a hand over his shoulder ....Need I tell you?

All stampedes are born of the monkey signal.

The great stampede at the cave-mouth was worth paying to see.

Hare was the first out. He went at a straight, most wonderful speed, ears flapped shut to drown out any possibly distracting sound, yet muttering: "Any creature who took his running lesson from God shall lose to cunning Man".

Monkey made it back to the tree-top. He has since sat there to peer at us.

Hyena scampered and kicked his hindlegs, in case Man had stuck to his tail. All the same, he warned his legs: "Any limb that breaks at this hour shall be abandoned to attend to itself", then a rude fart.

Snake, peeved that Man had taken herbs from him would try to slay him, slipped behind the cave, get his own back.

Quite at loss on the cause, Man took the heat off the fire, and ate a piece. The rest he stored, coming far a bite when he felt for. Not only had he become a cave dweller and eater of flesh, he also sought to right what went wrong.

He proceeded to hunt those who parted his company. After each kill he would flay the animal and hang the dried skin in the room, as a memento.
Gradually he built a pile of skins, until the idea of decorating the room with those skins entered his mind.
0
But was this man truly the only human inhabitant of the earth? Out hunting one morning, he spotted a form he took for a giant monkey. Carefully he stalked it. What was he to see? Behold, there stood the female of himself. He led her to the cave.

By the sprouting-green-time of the following year she bore them a male baby. (Now there was reason to give names). They named their son Ayamgo meaning `Stranger`.

Man carried on hunting to nourish his family. :Ayamgo grew to a fine child - active and alert. It was at this time that Snake took his revenge. That evening he slipped under Man and struck at his heel. The solace was he killed Snake, too, and kept the skin.

Were they truly the only humans on earth? Still smarting in the pain of *the* Snake's bite, he resolved to carry a last hunt and retire. One animal that had eluded him far too long was the Buffalo. Even in pain he searched for it.

Before sundown that day he spotted what he took to be a buffalo. Blinded by desire to act before the game may escape, he shot off his arrow. The arrow, instead of embedding, tore a long wound on the side of this animal. Out of his torn wound the foetus of a calf fell out. He had killed a she Buffalo in gestation.

So shocked, Man stood; looking. The foetus wriggled, then staggered to its feet, wobbled, began turning, to see her dying mother. Conveying a pathetic glance, she nevertheless sobbed "I shall avenge my mother" and then wobbled away.

Well, in the North, killing a gestating wild animal is far, far more calamitous than anything else.

What could he do now? He had orphaned a baby. The woman buffalo was dead. This was his last hunt.

He withdrew his arrow and flayed the cow. He discarded the body and went home to his safe family.

The cow's hide was the last of his collection of skins.

The few days following, his leg got worse, for it began to ooze pus. In a dream one night a cure was revealed to him; if only he could get the tail of a rhinoceros and with it keep flies away from the sore, he would heal.

He, hunter, could not move, downed by a sore leg. Rhinoceros lived miles across the river - the most dangerous and most solitary of all animals.

He called young Ayanga; and explained the trouble to him. The great blade that accompanied Man on all of his hunting expeditions but now in retirement, was plucked off its hanging hook. Ayamga must go and bring Rhinoceros' tail.

On his first day out, he came back promptly, but with Hare's tail. The second day out he brought Hyena`s tail. On the third day it was Leopard's tail. All these, Man took but sent him out again. This was where the collection of trophies began.

When this seemed to be going on endlessly, Man gave Ayamga a proper talking to. "Look, the one that brought you to the world. If you risk your life to save his life he would live again to save yours."

This time, Ayamga actually came across the river. He covered the width of a field from one side to another. About dusk he caught sight of two shining pebbles at a distance. He started towards there. From a stone-throw he could see whiffs of curling smoke. Close up, now, he couldn't mistake it for anything else. Before him, for sure, there slept the most dangerous animal of all.

Ayamga made an about turn. Today he had made the greatest effort yet only today he was going home empty handed. When he stumbled upon an
elephant the fear of shame gave him courage. He went behind Elephant and delivered a powerful swipe, making away with the tail. Though Elephant looked and saw he didn't come after him. It was this nonplussed attitude of Elephant that let contempt for huge animals creep into him.

early the next morning when he set out, caution thrown to the winds, he came directly and straight to Rhino's tail. Zap! Off it fell into his hand! Off he fled!

Great Rhino, so roused by a neverbefore experienced job, grunted, sniffed the air, and came in pursuit. Ayamga was already over the other side when the terrible one got to the bank. In no time he too crossed, but the gap was considerable.

Rhino`s stompings caused earth tremor and thunder. His nostril steam sparked lightning. Still running the boy sang a song of distress:

I cut Hare tail for father
Father wants Rhino tail

I cut Hyaena tail for father
Father wants Rhino tail
I cut Leopard tail for father
Father wants Rhino tail
I cut Lion tail for father
Father wants Rhino tail
I cut Hippo tail for father
Father wants Rhino tail
I cut Elephant tail for father
Father wants Rhino tail
Can father now hear Rhino?
Father, can you hear Rhino?
This is Rhino saying
Kalong Woolaam!
Kalong Woolaam!
Kalong Woolaam!
Damm Killoomm!
Damm Killoomm!
Heard Rhino Come?
Seen Rhino Flash?
felt Rhino Shake?
Kalong Woolaam!
Kalong Woolaam!
Kalong Woolaam!
Damm Killoomm!
Damm Killoomm!

Being far away like that, Man couldn't have heard the terrified words of his son. Fortunately that wasn't necessary. The clanging bells, the earth tremor, plus thunder and flashes, all told man what to prepare for.

Very quickly he got a fire built to inferno proportions. Once again he would use his spear, hopefully for the last time. He could imagine his son's situation so he felt the danger for himself as well. If the boy got torn up, abscess would waste him to death. But so long as the quakes and clangs kept closing in his hopes also went rising. his heart was out there with Ayamga. As a distinguished hunter he knew where to jay in wait.

Ayamga did prove himself the offspring of the greatest hunter ever to walk on earth. He burst into view and straight: into the cave. Rhino had actually narrowed the gap to an uncomfortable close. The instant Rhino burst up, Man proved himself to be the first human that pursued spearholding as an art. He sent a glowing spear flying and piercing tile Rhino's loins through. Death was instantaneous. Another skin had been added.

The potion ,proved the highest potent as an antibody. Ayamgo had made the mark of maturity.

It wasn't long until they heard that there were other humans on the earth. These people already had a market - not a market where people came to trade but a market where humans came to see other humans.

The first time Ayamgo visited this place he was extremely happy about the prospect of making acquaintainces. He paid three more visits on which time he met a most beautiful girl, beautiful beyond description - a beauty of a type that has never since been recorded in all human endeavour. Ayamgo being a fine son, sprout of hunter's stock, easily got her ensconsed with him.

The girl, wherever she came from, asked to go and know Ayamgo's home. Surfing on the crest of euphoria, son of hunter led his find to the cave.

It was one of those nice days known only in those good times. Man had retired from the active pursiut of his craft. These days he laid outside under a shed he had built of sticks, on the skin of the Rhino - that being the largest of his collections. Ayamga led his girl inside the cave.

The vivid decoration of the walls captivated the girl. She examined a pile of skins. Then she requests they go through them one by one:

Ayamgo, what skin is this?
Father ever killed a Hare
The Hare skin is that
Ayamgo, what skin is this?
Father ever killed a Hyena
The Hyena skin is that
Ayamgo, what skin is this?
Father ever killed a Leopard
The Leopard skin is that
Ayamgo, what skin is this?
Father ever killed a Lion
The Lion skin is that
Ayamgo, what skin is this?
Father ever killed a Snake
The Snake skin is that
Ayamgo, what skin is this?
Father ever killed a Hippo
The Hippo skin is that
Ayamgo, what skin is this?
Father ever killed a Roan
The Roan skin is that
Ayamgo, what skin is this?
Father ever killed a Buck
The Buck skin is that
Ayamgo, what skin is this?
Father ever killed a Zebra
The Zebra skin is that

Ayamgo, what skin is this?
Father ever killed a Giraffe
The Giraffe skin is that
Ayamgo, what skin is this?
Father ever killed a Antelope
The Antelope skin is that
Ayamgo, what skin is this?
Father ever killed a Buffalo
The Buffalo skin is that

When they had gone through the whole pile Ayamga noticed a mist of tears in the beauty's eyes. "Why are you sobbing?" he asked, endearingly.

"Oh no, just a strand of hair brushed my eyeball," she explained.

That was enough. She asked to be seen off home. He led her out, onto the road to a point where she said goodbye, promising to meet him at the market on the third day.

On the third day, Ayamga met her in the market. It was a repeat of their earlier encounter. Again he led her home and they went through the pile of skins. At the end she did shed tears. Then he saw her off.....

So on, on, it dragged on. When old hunter Man wasn't seeing any headway to his acquiring a beautiful daughter-in-law, be interfered. "What is going on between you and her?" he demanded. Ayamga gave his father the full account of their encounter, from the first day to date. No minor matter was left out be it a wry smile, a chortle, a laugh or a sneeze, he disclosed it in the magnitude of its exactitude. The old hunter delved deep into his memory. This time, when Ayamga had left for the market he entered the cave, and rearranged the pile of skins, putting the last on top. Back under the shed he came and lay an the Rhino skin.

On time, as expected, the first love-match of the earth came, hand in hand.. Inside the cave, the girl walked straight to the pile:

Ayamgo, what skin is this?
Father ever killed a Buffalo
The Buffalo skin is that

She didn't go further. Mist in her eyes formed teardrops. Still she explained that there were strands of hair in her eyes. The departure today was unusually early. He led her to the known point and bade goodbye.

For the second time, Man asked his son what had happened. Yet again Ayomga gave him the concise details. Old hunter knew what next to do.

The third day duly came. Ayomga went to the market and the pair came back. The moment they stepped inside the cave the old Hunter called out his son and hinted: "Son, today he alive and alert to danger. Plant yourself in a position to escape a second beast, like you did from Rhino." He added something else.....

Ayamga, what skin is this?
MY father killed YOUR mother
That`s YOUR mother`s skin.

Anyone who has ever stared into the eyes of a buffalo would know this. Those misty eyes on the beautiful face turned auburn and glowed!. She changed into a stout buffalo and tore after Ayamga who was already outside and fleeing.

Buffalo got nowhere when she reared up. Old hunter had his spear waiting. It drove through her ribs appearing at the other side. As she lay kicking to death, the beautiful face of the girl returned. She died, torso-beast head-girl. That is why:

Because I was there when all this happened, if I didn't let you know, how were you going to understand why our elders insist that if you met a girl and fancied her, you must first ask her name, ask her clan to know their taboo, verify this from another source, confirm for yourself that she is human by visiting her home, before you dream of bringing her to your home.

That is why, if it's not the craft of your clan, and if you haven't performed the rituals, you can't hunt a beast for how could you tell a gestating animal? How would you tell which animal was the spirit of a certain fetish.

The moral lessons in this story are various. What you might take to be a moral wouldn't be to a Bongo native.

Right now I imagine myself telling Africans that I have written a Bongo story on a piece of paper.

******

IS THERE ANYBODY OUT THERE?

by Graham Inglis

Frank Drake was a leading US astronomer for thirty years and was a front-runner in Humanity's fledgling attempts to make contact with hypothetical extra-terrestrial life-forms. (See Appendix for a brief biography.) Arguably he is most famous for having formulated a mathematical concept which for three decades has borne his name - the Drake Equation. This equation seeks to assess just how many contactable alien species are 'out there'.

Happily for most of us, this equation does not need any deep knowledge of mathematics. The difficulties arise not from wrestling with arithmetic but from general lack of knowledge. The component probabilities stretch across many fields, starting with astronomy and ranging through diverse areas including geology, chemistry, biology, history, anthropology, and politics - fields in which our understanding is woefully incomplete. To illustrate just how basic the *numerical* process is, consider a town which has 20 pubs or bars. If we are told that, as a national statistic, one pub in four serves cooked meals, then it doesn't take a mathematical genius to work out that we can hope that **five** pubs in that town offer hot food - we simply divide 20 by 4. Or, to be precise, we multiply 20 by the probability fraction of ¼. Drake's equation uses the same simple process.

Various versions of the Drake Equation exist. One runs thus:

'Aliens' = $N \times P \times H \times L \times I \times C \times S$

where N is the start figure and the following elements are the various probabilities that we multiply N by, to reduce it down to our target figure, "aliens".

N : the starting figure

As a numerical starting point - the equivalent to the 20 pubs in our hypothetical town - we can take the number of stars (or suns; the two words mean the same) in our own galaxy - the "Milky Way" galaxy - and, by elimination, end up with an assessment of the number of suns that could be hosts to intelligent (and hopefully communicative) species.

Opinions vary on just how many billions of stars there are in our Galaxy. 200 billion is often quoted by astronomers as a likely total. Others think that figure is too low, but we have to start somewhere. If we take 200 billion as our numerical launching pad, we can now start to shrink that number down - down to the number of those stars that are *possibly* home to contactable aliens - ie, technologically communicative lifeforms.

P : the fraction of stars that have planetary systems

Assuming that intelligent life must evolve on a planet is only the first of many assumptions the Drake Equation forces us to make. It's perhaps pessimistic: if such life can evolve on, say, a comet, then the possibilities for intelligent life immediately improve.

Many astronomers believe that more than 30% of the stars in our Galaxy have planetary systems. A not-unreasonable suggestion, as there's plenty of material out there in the galaxy - some of which hits Earth as meteors. One-third of the 200 billion stars in our galaxy yields a figure of around 66 billion planetary systems.

H : the fraction of those planetary systems that include "habitable" planets

There may well be 66 billion solar systems out there but if each system merely contained planets like Jupiter (gas) or Pluto (barren rock), then the chances of life as we know it (pardon the cliché) forming there do not look promising.

We presumably need to look for a solid planet at least Earth-size (smaller ones, like Mercury, have insufficient gravity to sustain a substantial atmosphere) and which orbits at a comfortable and reasonably consistent distance from its star - not too close and not too far away, so as to avoid excesses of heat or cold. Astronomers differ widely on how likely all this is - estimates range from an optimistic 50% to a pessimistic hardly-ever-happens assessment.

Many stars exist as close pairs or multiples and many astronomers question whether a stable planetary system can exist in a double or multiple star system. If a stable system formed, it is likely that such planets would suffer extremes of heat and cold as the distances between it and either of its two suns varied.

Steering a middle course between the optimists and the pessimists, a probability of 1 in 1000 seems reasonable. 66 billion divided by 1000 yields 66 million planets...

L : the fraction of habitable planets where life arises

The first factor to be considered is just what we mean by 'life'. Of course, whole books have been written on the question. A *very* short definition might be: a biochemically cohesive entity that interacts with its environment.

Current stellar evolution theory suggests the early universe lacked rocky planets because only helium and hydrogen was around, and this mix eventually gave birth to the heavier elements (like carbon, iron, silver, etc) in supernovae - atomic fusion in exploding suns. It is estimated that our own solar system formed 4.7 billion years ago from material ejected from earlier supernovae.

It is generally believed that the formation of life requires conditions such as tidal rock-pools and atmospheric lightning - and an appropriate mix of chemicals in the water and air.

The probability of enough life-forming chemicals accumulating in the right conditions will obviously increase over time. Current biological and chemical knowledge seems to indicate that life is still a highly *unlikely* thing to occur, even if the appropriate conditions are present and billions of years elapse. However, scientists disagree markedly on this question: some believe it's almost a certainty (given time) whilst others think such an event is vanishingly rare.

Some biologists regard it as overwhelmingly probable that any form of proto-life that comes into being will never evolve into 'life' simply because there's no reason why it should; there's no impetus (or watchmaker).

However exacting the conditions for the *emergence* of life may be, it seems clear that, once life *has* started, it can survive in surprisingly hostile conditions. In the last decade or so we have discovered that life-forms on this planet can be more tenacious than ever previously imagined. Some thrive around hot springs deep under the sea where virtually no sunlight penetrates. Some bacteria can survive years of sub-zero temperatures, or thrive in radioactive areas, for example.

Life - and more particularly, consciousness - seems to depend upon hectic activity and change at the molecular level. On planet Earth, the individual cells of living organisms all contain a microscopic architecture of molecules - an intrinsically unstable architecture that is constantly renewing itself. The heart of this biological architecture is the twisted and looped DNA molecule - the double helix - found in the cell nucleus.

But is this intricate and specialised structure the only route to biological life?

"I'm a doctor, not a brick-layer," grumbled Star Trek's Dr McCoy, as he 'repaired' a wounded silicon-based lifeform with a handful of cement. The creature, rather resembling a large and bad-tempered carpet, and known as a 'Horta', featured in an episode called "The Devil in the Dark". This silicon-based creature is but one of many science fiction speculations as to the possibility of a lifeform based upon "alternative" biochemical processes.

It is an intriguing idea, but could it actually be possible? Are there other chemical pathways along which life could develop?

Our DNA molecule is complex - necessarily so, since it contains the coded infor-mation needed for control of and repair of the many cell structures outside the nucleus. DNA issues instructions; enzymes carry out the actual biochemical chores.

Life as we understand it depends upon various gradients. Areas of high and low electrical potential enable movement of charged ions, in the same way that a stone will readily roll downhill. Areas of strong and weak concentrations of a substance can encourage osmotic migration of material across a semi-permeable membrane. Without decay and replication within a cell, the overall environment within that cell would tend to average out - and it would cease to live. Furthermore, a lack of replication errors would make evolution through natural rejection far less likely. It seems that life requires continual and dynamic conflict at the molecular level before it can flourish.

Silicon and carbon are in the same group within the periodic table of the elements - but then, so are the decidedly metallic elements tin and lead. Silicon (chemical symbol Si) is a hard, dark gray non-metallic element, which comprises 28% of the earth's crust and is second only to oxygen in abundance upon this planet.

While there are no known silicon-based lifeforms on Earth, there are animals that have skeletons made not of calcium compounds but of silicon dioxide (silica). Known as radiolarians, these are one-celled animals, living in the depths of the oceans, that have taken silica from their external environment and have built it up as a protective outer wall - a primitive shell that rather resembles a snowflake in appearance. The practice of utilising Earth's silicon stocks has not caught on in evolution generally, however: calcium compounds just seem easier to handle than silicon ones when it comes to building skeletal structures.

The phrase 'silicon-based lifeform' is generally taken to mean one where silicon acts as a substitute for carbon in the basic building blocks of life. On the face of it, this possibility seems plausible, since carbon and silicon are fairly similar, chemically speaking. However, their compounds are likely to be very different on the physical level. For instance, compare the dioxides of silicon and carbon.

CO_2 and SiO_2 are chemical cousins, both being stable compounds generally unreactive with other substances. Their physical characteristics, however, could hardly be more different, since CO_2 is a gas at room temperature and SiO_2 (quartz - the main component of glass and sand) is a fairly brittle crystalline solid.

Carbon-based lifeforms depend upon carbon chains for their replication. Unfortunately for any would-be silicon-based lifeform, silicon does not 'chain' so readily. The Si-Si (silicon-to-silicon) bond energy is nearly half that of the corresponding C-C carbon bond energy. Instead, silicon atoms tend to crystallise, ie cross-connect in a sprawling lattice fashion. Carbon can do this too, of course - but less readily: which is why diamonds (crystallised carbon) are comparatively rare.

Crystalline life, such as that described in Michael Creighton's book *The Andromeda Strain*, should not be too hastily ruled out. The university chemist Graham Cairns-Smith suggests that mineral crystals are quite capable of replicating and evolving. Crystals obviously can grow, since they tend to build up in a consistently-repeating pattern. They can also multiply if they shear along their natural crystal planes. Furthermore, crystal surfaces can act as efficient facilitators for many chemical reactions, including some that involve organic chemicals.

Cairns-Smith has made the intriguing suggestion that, approximately 4 billion years ago, populations of replicating and evolving crystals on Earth began to make various organic chemicals for their own purposes, perhaps to use as binding agents. These organic molecules then independently became capable of replication - and, Cairns-Smith suggests, evolved into the organic life that now abounds upon Earth.

Are we descended from a type of Frankenstein's Monster that ran out of control of its creators? Could it be that we are the result of a crystalline evolutionary experiment that went wrong? This is worth considering, next time you are walking on a sandy beach!

In general terms, however, it seems that a 'typical' crystalline organism would not actually do very much, at least by our standards. It would be a fairly loose collection of interacting crystals whose compositions and boundaries were indistinct and long-term intentions (if any) even more indistinct. Intelligence and purpose in such an entity seems rather unlikely to develop beyond the capabilities of, say, moss.

Earth life at the cellular level is usually carried out in a solvent in which various molecules and ions are dissolved or suspended. This substance - water - is the most widely-effective natural solvent known. It is for this reason that many biologists make the assumption that planetary life could not emerge without water: at the moment we do not know of any other suitable and widely-available solvents.

Assuming for the moment that all life requires change and replication at the molecular level, doing it within a solvent is undoubtedly effective, as our own existence demonstrates. But in many ways, the underlying principle that seems to apply to life is simply the requirement of an open but differentiated system through which energy and certain materials flow.

Forms of microscopic life have been found on this planet in the vicinity of volcanoes and in the Antarctic that seem to thrive with very little water. Recently, a whole biotope was discovered in a network of caves beneath the country we know today as Romania. The creatures that live there have evolved independently for thousands of years, and thrive under conditions which would kill most other animals and plants. Some creatures have evolved to live in the sulphurous hot springs of volcanic areas in what, to most creatures, is a poisonous chemical soup.

Even these, however are not the strangest and most isolated creatures to be found on the globe. In 1977 the US Navy submarine Alvin was exploring hydrothermic vents on the floor of the Pacific Ocean near the Galapagos Islands when the crew were amazed to find a whole ecosystem based around them. As Dr Karl Shuker explained in his 1993 book *The Lost Ark - New and Rediscovered Animals of the 20th Century*:

"The vent world's discovery, however, unveiled the operation of a radically new, autonomous energy chain, one that was completely independent of the sun. Instead of sunlight, the energy source being used was chemical energy - released during the combination of sulphates with hydrogen to form hydrogen sulphide (....) The released energy was being utilised by sulphur-oxidising bacteria to manufacture their own food. In turn, they were being eaten by larger life-forms, which were themselves being preyed upon by others, and so on..."

Other oceanic 'vent worlds' have since been discovered, and the chemical balance of some of these oases of life seem to be even stranger. These discoveries indicate that life is more diverse and hardy than many science fiction enthusiasts of earlier years would have dared to imagine, and indicate that sunlight and the solvent known as water might not be as indispensable as we once thought.

Interacting materials can be distributed in ways other than being dissolved in a solvent. Tobacco smoke hangs in the air, but the air is not a solvent - the particles are in suspension and eventually settle out. A creature consisting solely of gas could, in theory, have much internal chemistry going on - but the absence of internal structure and any semi-permeable membranes would seem to preclude the necessary condition of ongoing chemical conflict mentioned earlier. A uniformly-mixed zone of gas and suspended particles would presumably not have the capacity for internally-initiated reactions, since its internal entropy would be at a maximum.

However, creatures in the form of balloons containing gas have been proposed as candidates for life on Jupiter. Carl Sagan, in his book Cosmos, described an intriguing colony of 'gas bag' creatures drifting among the ammonia-laced clouds.

The astronomers Fred Hoyle and Chandra Wickramasinghe have suggested that life evolved in space - and that these aliens often reach Earth. The organisms that they have postulated are not malevolent Greys who commit acts of criminal damage in farmers' fields and abduct female hippies, however, but are micro-organisms riding on dust particles propelled by light from the stars. Earth was 'seeded' with life, it is suggested, as would have been other planets. If the Martian meteorite* ALH 84OO1, recovered from Antarctica, is shown to contain the remnants of micro-organisms from Mars, this would be an interesting piece of evidence in support of the theory.

Supporters of the theory still have to overcome the difficulty in describing a detailed possible mechanism for such emergence of life.

One of the main problems humans have when contemplating the universe at large is the difficulty in avoiding being parochial. For instance, when we describe iron as a solid, we usually forget to add the important words "at room temperature" - where the 'room' referred to is a room on Earth maintained at a temperature to our own liking. At temperatures some hundreds of degrees above 'room temperature' silicon, while still a solid, becomes less prone to forming lattice structures.

While the chemistry of silicon at such temperatures has not been investigated in any great depth, it seems that its chaining potential (such as it is) is relatively unimpaired. On a suitably hot planet, then, short silicon - based chains could perhaps be utilised by a lifeform as an equivalent to DNA. The chain would not be nearly as long as its carbon counterpart - but in certain situations this might not matter. So long as different sections of the genetic memory store do not interfere with each other, they don't necessarily have to be connected end-to-end to achieve their purpose. Lifeforms emerging on this 'suitably hot planet' would obviously be unable to use water as a life-sustaining solvent, since water in such an environment would be in the form of steam.

The long-term storing of information seems vital, since life seems intrinsically dependent upon being in a state of dynamic near-disaster, and something presumably has to keep the big picture in mind. The precise nature of the storage template seems unimportant, though: so long as it does the job.

Science fiction stories about silicon-based lifeforms generally portray the creatures as boulder-chomping toughs. The 'Horta' which starred in the Star Trek episode mentioned earlier was a blob-like creature that burrowed through rock and brought consternation to the mining community of that planet. A far simpler method of obtaining growth-enabling silicon would be to let plants break down the rocks and then just live off the plants - in much the same way that animals on Earth cannot obtain nitrogen from the air to formulate new supplies of nucleic acids, but instead depend (directly or indirectly) on plants that can "fix" atmospheric nitrogen. It is possible that silicon-based animals, far from being boulder-pulverises, would actually be relaxed vegetarians.... but that, of course, wouldn't make half as exciting a television programme!

Some biologists believe that, given the appropriate conditions, the development of life is a virtual certainty. The more we look at the matter, the more diverse the possibilities appear.

I: the fraction of lifeforms that develop intelligence

Let's say that, in some instances, life splutters into being and manages to achieve some means of self-replication. How likely is it that this life will ever develop and become what we like to call 'advanced'?

Some biologists think intelligence is an inevitable consequence of evolution. How-ever, such a supposition is little more than disguised faith. Those who confuse wishful thinking and science effectively are making a religion - and then usually seek to deny that they are doing this. Scientists often confuse theory and fact, because there is generally no real meaningful difference between the two. One man's fact is another man's irrelevance and a third man's article of faith.

An evolutionary change (eg an infra-red detector spot taking a step towards evolving into a light-detecting eye) might well be a good idea, but evolution isn't driven by good intentions. A change that facilitates future development won't, at the time, necessarily be advantageous to the organism concerned and the change could, therefore, fizzle out.

An enduring scientific and science fiction speculation is to wonder if the dinosaurs would have ever developed intelligence as we know it, had they not been wiped out by the Cretaceous Event 65 million years ago.

One can imagine millions of little biochemical variations fizzling and spluttering in a species, most having no effect one way or the other on survival ability; and some variations having a cumulative effect and eventually causing individual representatives of the species to either lose vital ground or to forge ahead in the survival race. Competition for resources can bring out the 'best' in a species, so far as future survival is concerned. I think that the dinosaurs, if they had been allowed to survive, *may* have developed intelligence - but most probably they wouldn't have.

Alignment with the pessimists seems indicated: given a few tens of thousand life-bearing planets, intelligence would perhaps only arise once or twice. Perhaps a probability of 1-in-50,000 again?

We started with about 200 billion stars and immediately whittled that down by one-third, yielding only 66 billion stars left in the running. Multiplying that number by 1/50,000 and then by another 1/50,000 leaves only 26 stars in the whole galaxy theoretically "in the running" for developing a communicative technology.

C : the fraction of intelligent species that develop technological communication

Given that intelligence arises, will technology and a desire for communication follow? The latter probably goes hand-in-hand with intelligence - such a race would have more ideas to discuss - but technology needn't. Human technology relates mainly to changing our environment, and changing the skeletal integrity of our enemies! The problem here, of course, is that we are forming theories based on a sample size of 1, ie ourselves. *We* developed technology so surely all intelligent lifeforms will...?

However, many anthropologists believe that our desire for technology (and war) is a symptom of an imbalance between our intellect and our ancestral instincts - we are able but not wise. A wise and mature race might develop a rather more holistic and Gaia-based society than ours. Erring on the side of caution again, and saying only 1 in 20 might develop communicative technology, will whittle the 26 planets now left in the running down to between one and two - and many people believe (for religious reasons) that just one race is what a God or Gods indeed did create: us.

S : the fraction of communicative civilisations that survive.

If the Drake Equation is evaluated in an optimistic frame of mind, one can postulate thousands of communicative civilisations having developed.

The species longevity question is a particularly problematic one. Might a fledgling civilisation destroy itself before maturing? Can an emerging civilisation survive the discovery of atomic power? Is a "mature" civilisation capable of violence? Answering these questions involves straying way outside the bounds of conventional science, and delving into politics, sociology and psychology - areas into which your average scientist never dares tread.

Some astronomers argue that it's safe for us to try to contact advanced civilisations, because they must be benign - or they wouldn't have survived. To claim certainty on this point is surely absurd. Is it impossible that a race could be internally benign, rabid so to speak, while being rabidly xenophobic when it comes to meeting outsiders? Surely not.

Conversely, surely not all Galactic species will grow to be as politically moribund and environmentally destructive as Humanity has demonstrated itself to be.

GIGO?

Our ignorance in each applicable field is what makes evaluating the equation so fascinating to some people (almost anybody's guess is equally valid) and so irritating to those who dislike the sheer fuzziness of it all. Somebody once said that the Drake Equation proves nothing except how ignorant humans are. However, as our knowledge of astronomy, biochemistry and other sciences advances, year on year, the more accurate our evaluation of the Equation can become. Therefore, its conclusions become less and less problematic as we advance our understanding in the relevant fields: we can plug new facts into the analysis and thus refine our conclusions.

With each new discovery we make, and with each new piece of data that is available, we can upgrade our evaluation of Drake's challenging equation.

At the time of writing, you can try inputting different values into the Drake Equation at the following Internet webpage:

http://www.activemind.com/Mysterious/Topics/SETI/drake_equation.html

----

* The ALH 84001 meteorite is discussed by the author in *Animals & Men* magazine issue 11, p.27

APPENDIX

Frank Drake was born in Chicago on May 28, 1930.

He became interested in science and he and his friends would spend hours experimenting with motors, radios, and chemistry sets. As his understanding of astronomy and the actual size of the universe grew he began to wonder about the possibility of the existence of other planets and life on those planets. The idea seemed reasonable to him. However, because of the religious convictions of his parents and teachers he never felt comfortable bringing up the subject of extraterrestrial life.

After high school Drake enrolled at Cornell University to study electronics. It was here that he fell in love with astronomy.

In 1951, during his junior year, he attended a lecture by Otto Struve, one of the world's pre-eminent astrophysicists. Towards the end of a lecture Struve showed that there was mounting evidence that planetary systems had most likely formed around half of the stars in the galaxy. Struve went on to state that life could certainly exist on some of those planets. Finally, Drake had found someone who shared his ideas. After college he spent the next three years with the Navy to repay his scholarship. Thanks to his electronics degree he ended up as the electronics officer on the USS Albany where he gained invaluable experience operating and fixing the latest high-tech electronic equipment.

When his Navy tour ended, Drake went to Harvard graduate school and studied radio astronomy. Because of his electronics experience in the Navy he was a natural fit because the radio astronomy equipment was constantly in need of tweaking and repair. Drake got hooked on radio astronomy and never looked back.

Upon finishing graduate school in 1958 he got a position at a radio telescope in West Virginia. It was here in 1960 that the first "search for extraterrestrial intelligence" (SETI) took place. Named Project Ozma by Drake, they scanned the stars Tau Ceti and Epsilon Eridani for two weeks, for signs of radio transmissions.

In 1961 Drake helped organized the first SETI conference. The three day meeting was a small gathering of a dozen or so scientists who had shown an interest in SETI. It was in preparation for this conference that Drake came up with the now famous Drake Equation, as a way to focus on the factors which determine how many intelligent, communicating civilisations there might be in our galaxy.

It took the form

$N = N^* f_p n_e f_l f_i f_c f_L$

and aimed to focus the conference attendees' attention on the key questions that needed to be answered in order to determine the chances of SETI's success.

Drake was to remain active in US and international astronomy for over three decades - including directorship of the Arecibo Observatory in Puerto Rico and as sometimes-consultant to NASA.

****

1998 - A YEAR IN THE LIFE OF THE CENTRE FOR FORTEAN ZOOLOGY

Our seventh year of existence was in many ways our most momentous. It started well when in January Graham Inglis and I embarked on a mammoth trek across Central America, visiting Puerto Rico, Mexico and Florida in search of the fabled chupacabra. As regular readers will know, we didn't find it, but we *did* find something which to us was pretty exciting - what looks as if it is a new species of lamprey!

When we returned to the UK we found that work had piled up whilst we were gone. Within weeks of our arrival back in Exeter we were embarked on a long search for `The Beast of Haldon` - a mysterious creature (vaguely described as cat-like) that has been reported for several years digging through the graves at a local pet cemetery.

Our friend and colleague Richard Freeman joined Graham and me on our search. By the time he had finished his degree course at Leeds University in July he had spent so much of the year staying with us that we all decided that it would be a pretty good idea of he were to come down and join the team on a permanent basis.

Much of the rest of the year was spent doing routine stuff like writing books and magazine articles, keeping the administrative machine of the CFZ in some kind of order and appearing on low budget TV chat shows but during the final five months of the year we continued our field investigations into The Beast of Haldon, travelled to London on many occasions to work on various TV projects, filmed the semi legendary `Walking of the Black Dog` procession in mid Devon and, oh yes, we started making a full length movie based on my 1997 book *The Owlman and Others*. So all in all a reasonably good year..

Jonathan Downes
The Centre for Fortean Zoology, Exeter,
March 1999.

THE CENTRE FOR FORTEAN ZOOLOGY

So, what is the Centre for Fortean Zoology?

We are a non profit-making organisation founded in 1992 with the aim of being a clearing house for information, and coordinating research into mystery animals around the world. We also study out of place animals, rare and aberrant animal behaviour, and Zooform Phenomena; – little-understood "things" that appear to be animals, but which are in fact nothing of the sort, and not even alive (at least in the way we understand the term).

Why should I join the Centre for Fortean Zoology?

Not only are we the biggest organisation of our type in the world but - or so we like to think - we are the best. We are certainly the only truly global Cryptozoological research organisation, and we carry out our investigations using a strictly scientific set of guidelines. We are expanding all the time and looking to recruit new members to help us in our research into mysterious animals and strange creatures across the globe. Why should you join us? Because, if you are genuinely interested in trying to solve the last great mysteries of Mother Nature, there is nobody better than us with whom to do it.

What do I get if I join the Centre for Fortean Zoology?

For £12 a year, you get a four-issue subscription to our journal *Animals & Men*. Each issue contains 60 pages packed with news, articles, letters, research papers, field reports, and even a gossip column! The magazine is A5 in format with a full colour cover. You also have access to one of the world's largest collections of resource material dealing with cryptozoology and allied disciplines, and people from the CFZ membership regularly take part in fieldwork and expeditions around the world.

How is the Centre for Fortean Zoology organized?

The CFZ is managed by a three-man board of trustees, with a non-profit making trust registered with HM Government Stamp Office. The board of trustees is supported by a Permanent Directorate of full and part-time staff, and advised by a Consultancy Board of specialists - many of whom who are world-renowned experts in their particular field. We have regional representatives across the UK, the USA, and many other parts of the world, and are affiliated with other organisations whose aims and protocols mirror our own.

I am new to the subject, and although I am interested I have little practical knowledge. I don't want to feel out of my depth. What should I do?

Don't worry. We were *all* beginners once. You'll find that the people at the CFZ are friendly and approachable. We have a thriving forum on the website which is the hub of an ever-growing electronic community. You will soon find your feet. Many members of the CFZ Permanent Directorate started off as ordinary members, and now work full time chasing monsters around the world.

I have an idea for a project which isn't on your website. What do I do?

Write to us, e-mail us, or telephone us. The list of future projects on the website is not exhaustive. If you have a good idea for an investigation, please tell us. We may well be able to help.

How do I go on an expedition?

We are always looking for volunteers to join us. If you see a project that interests you, do not hesitate to get in touch with us. Under certain circumstances we can help provide funding for your trip. If you look on the future projects section of the website, you can see some of the projects that we have pencilled in for the next few years.

In 2003 and 2004 we sent three-man expeditions to Sumatra looking for Orang-Pendek - a semi-legendary bipedal ape. The same three went to Mongolia in 2005. All three members started off merely subscribers to the CFZ magazine.

Next time it could be you!

Project Kerinci, Sumatra - 2003
In search of the bipedal ape Orang Pendek

How is the Centre for Fortean Zoology funded?

We have no magic sources of income. All our funds come from donations, membership fees, works that we do for TV, radio or magazines, and sales of our publications and merchandise. We are always looking for corporate sponsorship, and other sources of revenue. If you have any ideas for fund-raising please let us know. However, unlike other cryptozoological organisations in the past, we do not live in an intellectual ivory tower. We are not afraid to get our hands dirty, and furthermore we are not one of those organisations where the membership have to raise money so that a privileged few can go on expensive foreign trips. Our research teams both in the UK and abroad, consist of a mixture of experienced and inexperienced personnel. We are truly a community, and work on the premise that the benefits of CFZ membership are open to all.

What do you do with the data you gather from your investigations and expeditions?

Reports of our investigations are published on our website as soon as they are available. Preliminary reports are posted within days of the project finishing.

Each year we publish a 200 page yearbook containing research papers and expedition reports too long to be printed in the journal. We freely circulate our information to anybody who asks for it.

Is the CFZ community purely an electronic one?

No. Each year since 2000 we have held our annual convention - the *Weird Weekend* - in Exeter. It is three days of lectures, workshops, and excursions. But most importantly it is a chance for members of the CFZ to meet each other, and to talk with the members of the permanent directorate in a relaxed and informal setting and preferably with a pint of beer in one hand. Starting this year-18-20 August 2006 - the *Weird Weekend* will be bigger and better and held in the idyllic rural location of Woolsery in North Devon.

We are hoping to start up some regional groups in both the UK and the US which will have regular meetings, work together on research projects, and maybe have a mini convention of their own.

Since relocating to North Devon in 2005 we have become ever more closely involved with other community organisations, and we hope that this trend will continue. We also work closely with Police Forces across the UK as consultants for animal mutilation cases, and during 2006 we intend to forge closer links with the coastguard and other community services. We want to work closely with those who regularly travel into the Bristol Channel, so that if the recent trend of exotic animal visitors to our coastal waters continues, we can be out there as soon as possible.

We are building a Visitor's Centre in rural North Devon. This will not be open to the general public, but will provide a museum, a library and an educational resource for our members (currently over 400) across the globe. We are also planning a youth organisation which will involve children and young people in our activities.

Apart from having been the only Fortean Zoological organisation in the world to have consistently published material on all aspects of the subject for over a decade, we have achieved the following concrete results:

- Disproved the myth relating to the headless so-called sea-serpent carcass of Durgan beach in Cornwall 1975
- Disproved the story of the 1988 puma skull of Lustleigh Cleave
- Carried out the only in-depth research ever into mythos of the Cornish Owlma
- Made the first records of a tropical species of lamprey
- Made the first records of a luminous cave gnat larva in Thailand.
- Discovered a possible new species of British mammal - The Beech Marten.
- In 1994-6 carried out the first archival fortean zoological survey of Hong Kong.
- In the year 2000, CFZ theories where confirmed when an entirely new species of lizard was found resident in Britain.
- Identified the monster of Martin Mere in Lancashire as a giant wels catfish
- Expanded the known range of Armitage's skink in the Gambia by 80%
- Obtained photographic evidence of the remains of Europe's largest known pike
- Carried out the first ever in-depth study of the *ninki-nanka*
- Carried out the first attempt to breed Puerto Rican cave snails in captivity
- Were the first European explorers to visit the `lost valley` in Sumatra

EXPEDITIONS & INVESTIGATIOINS TO DATE INCLUDE

- 1998 Puerto Rico, Florida, Mexico *(Chupacabras)*
- 1999 Nevada *(Bigfoot)*
- 2000 Thailand *(Giant snakes called nagas)*
- 2002 Martin Mere *(Giant catfish)*
- 2002 Cleveland *(Wallaby mutilation)*
- 2003 Bolam Lake *(BHM Reports)*
- 2003 Sumatra *(Orang Pendek)*
- 2003 Texas *(Bigfoot; giant snapping turtles)*
- 2004 Sumatra *(Orang Pendek; cigau, a sabre-toothed cat)*
- 2004 Illinois *(Black panthers; cicada swarm)*
- 2004 Texas *(Mystery blue dog)*
- 2004 Puerto Rico *(Chupacabras; carnivorous cave snails)*
- 2005 Belize *(Affiliate expedition for hairy dwarfs)*
- 2005 Mongolia *(Allghoi Khorkhoi aka Mongolian death worm)*
- 2006 Gambia *(Gambo - Gambian sea monster , Ninki Nanka and Armitage s skink*
- 2006 Llangorse Lake *(Giant pike, giant eels)*
- 2006 Windermere *(Giant eels)*
- 2007 Coniston Water *(Giant eels)*
- 2007 Guyana *(Giant anaconda, didi, water tiger)*

To apply for a <u>FREE</u> information pack about the organisation and details of how to join, plus information on current and future projects, expeditions and events.

Send a stamped and addressed envelope to:

**THE CENTRE FOR FORTEAN ZOOLOGY
MYRTLE COTTAGE, WOOLSERY,
BIDEFORD, NORTH DEVON
EX39 5QR.**

or alternatively visit our website at:
www.cfz.org.uk

Other books available from
CFZ PRESS

THE OWLMAN AND OTHERS - 30th Anniversary Edition
Jonathan Downes - ISBN 978-1-905723-02-7

£14.99

EASTER 1976 - Two young girls playing in the churchyard of Mawnan Old Church in southern Cornwall were frightened by what they described as a "nasty bird-man". A series of sightings that has continued to the present day. These grotesque and frightening episodes have fascinated researchers for three decades now, and one man has spent years collecting all the available evidence into a book. To mark the 30th anniversary of these sightings, Jonathan Downes has published a special edition of his book.

DRAGONS - More than a myth?
Richard Freeman - ISBN 0-9512872-9-X

£14.99

First scientific look at dragons since 1884. It looks at dragon legends worldwide, and examines modern sightings of dragon-like creatures, as well as some of the more esoteric theories surrounding dragonkind.

Dragons are discussed from a folkloric, historical and cryptozoological perspective, and Richard Freeman concludes that: "When your parents told you that dragons don't exist - they lied!"

MONSTER HUNTER
Jonathan Downes - ISBN 0-9512872-7-3

£14.99

Jonathan Downes' long-awaited autobiography, *Monster Hunter*...

Written with refreshing candour, it is the extraordinary story of an extraordinary life, in which the author crosses paths with wizards, rock stars, terrorists, and a bewildering array of mythical and not so mythical monsters, and still just about manages to emerge with his sanity intact.......

MONSTER OF THE MERE
Jonathan Downes - ISBN 0-9512872-2-2

£12.50

It all starts on Valentine's Day 2002 when a Lancashire newspaper announces that "Something" has been attacking swans at a nature reserve in Lancashire. Eyewitnesses have reported that a giant unknown creature has been dragging fully grown swans beneath the water at Martin Mere. An intrepid team from the Exeter based Centre for Fortean Zoology, led by the author, make two trips – each of a week – to the lake and its surrounding marshlands. During their investigations they uncover a thrilling and complex web of historical fact and fancy, quasi Fortean occurrences, strange animals and even human sacrifice.

**CFZ PRESS, MYRTLE COTTAGE,
WOOLFARDISWORTHY BIDEFORD,
NORTH DEVON, EX39 5QR
w w w . c f z . o r g . u k**

Other books available from
CFZ PRESS

ONLY FOOLS AND GOATSUCKERS
Jonathan Downes - ISBN 0-9512872-3-0

£12.50

In January and February 1998 Jonathan Downes and Graham Inglis of the Centre for Fortean Zoology spent three and a half weeks in Puerto Rico, Mexico and Florida, accompanied by a film crew from UK Channel 4 TV. Their aim was to make a documentary about the terrifying chupacabra - a vampiric creature that exists somewhere in the grey area between folklore and reality. This remarkable book tells the gripping, sometimes scary, and often hilariously funny story of how the boys from the CFZ did their best to subvert the medium of contemporary TV documentary making and actually do their job.

WHILE THE CAT'S AWAY
Chris Moiser - ISBN: 0-9512872-1-4

£7.99

Over the past thirty years or so there have been numerous sightings of large exotic cats, including black leopards, pumas and lynx, in the South West of England. Former Rhodesian soldier Sam McCall moved to North Devon and became a farmer and pub owner when Rhodesia became Zimbabwe in 1980. Over the years despite many of his pub regulars having seen the "Beast of Exmoor" Sam wasn't at all sure that it existed. Then a series of happenings made him change his mind. Chris Moiser—a zoologist—is well known for his research into the mystery cats of the westcountry. This is his first novel.

CFZ EXPEDITION REPORT 2006 - GAMBIA
ISBN 1905723032

£12.50

In July 2006, The J.T.Downes memorial Gambia Expedition - a six-person team - Chris Moiser, Richard Freeman, Chris Clarke, Oll Lewis, Lisa Dowley and Suzi Marsh went to the Gambia, West Africa. They went in search of a dragon-like creature, known to the natives as `Ninki Nanka`, which has terrorized the tiny African state for generations, and has reportedly killed people as recently as the 1990s. They also went to dig up part of a beach where an amateur naturalist claims to have buried the carcass of a mysterious fifteen foot sea monster named 'Gambo', and they sought to find the Armitage's Skink (*Chalcides armitagei*) - a tiny lizard first described in 1922 and only rediscovered in 1989. Here, for the first time, is their story.... With an forward by Dr. Karl Shuker and introduction by Jonathan Downes.

BIG CATS IN BRITAIN YEARBOOK 2006
Edited by Mark Fraser - ISBN 978-1905723-01-0

£10.00

Big cats are said to roam the British Isles and Ireland even now as you are sitting and reading this. People from all walks of life encounter these mysterious felines on a daily basis in every nook and cranny of these two countries. Most are jet-black, some are white, some are brown, in fact big cats of every description and colour are seen by some unsuspecting person while on his or her daily business. 'Big Cats in Britain' are the largest and most active group in the British Isles and Ireland This is their first book. It contains a run-down of every known big cat sighting in the UK during 2005, together with essays by various luminaries of the British big cat research community which place the phenomenon into scientific, cultural, and historical perspective.

CFZ PRESS, MYRTLE COTTAGE, WOOLSERY, BIDEFORD, NORTH DEVON, EX39 5QR
w w w . c f z . o r g . u k

Other books available from
CFZ PRESS

THE SMALLER MYSTERY CARNIVORES OF THE WESTCOUNTRY
Jonathan Downes - ISBN 978-1-905723-05-8

£7.99

Although much has been written in recent years about the mystery big cats which have been reported stalking Westcountry moorlands, little has been written on the subject of the smaller British mystery carnivores. This unique book redresses the balance and examines the current status in the Westcountry of three species thought to be extinct: the Wildcat, the Pine Marten and the Polecat, finding that the truth is far more exciting than the currently held scientific dogma. This book also uncovers evidence suggesting that even more exotic species of small mammal may lurk hitherto unsuspected in the countryside of Devon, Cornwall, Somerset and Dorset.

THE BLACKDOWN MYSTERY
Jonathan Downes - ISBN 978-1-905723-00-3

£7.99

Intrepid members of the CFZ are up to the challenge, and manage to entangle themselves thoroughly in the bizarre trappings of this case. This is the soft underbelly of ufology, rife with unsavoury characters, plenty of drugs and booze." That sums it up quite well, we think. A new edition of the classic 1999 book by legendary fortean author Jonathan Downes. In this remarkable book, Jon weaves a complex tale of conspiracy, anti-conspiracy, quasi-conspiracy and downright lies surrounding an air-crash and alleged UFO incident in Somerset during 1996. However the story is much stranger than that. This excellent and amusing book lifts the lid off much of contemporary forteana and explains far more than it initially promises.

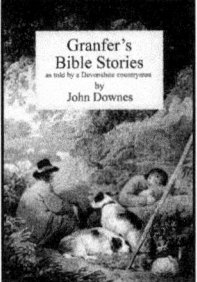

GRANFER'S BIBLE STORIES
John Downes - ISBN 0-9512872-8-1

£7.99

Bible stories in the Devonshire vernacular, each story being told by an old Devon Grandfather - 'Granfer'. These stories are now collected together in a remarkable book presenting selected parts of the Bible as one more-or-less continuous tale in short 'bite sized' stories intended for dipping into or even for bed-time reading. `Granfer` treats the biblical characters as if they were simple country folk living in the next village. Many of the stories are treated with a degree of bucolic humour and kindly irreverence, which not only gives the reader an opportunity to re-evaluate familiar tales in a new light, but do so in both an entertaining and a spiritually uplifting manner.

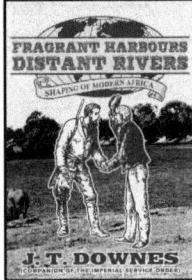

FRAGRANT HARBOURS DISTANT RIVERS
John Downes - ISBN 0-9512872-5-7

£12.50

Many excellent books have been written about Africa during the second half of the 19th Century, but this one is unique in that it presents the stories of a dozen different people, whose interlinked lives and achievements have as many nuances as any contemporary soap opera. It explains how the events in China and Hong Kong which surrounded the Opium Wars, intimately effected the events in Africa which take up the majority of this book. The author served in the Colonial Service in Nigeria and Hong Kong, during which he found himself following in the footsteps of one of the main characters in this book; Frederick Lugard – the architect of modern Nigeria.

**CFZ PRESS, MYRTLE COTTAGE,
WOOLFARDISWORTHY BIDEFORD,
NORTH DEVON, EX39 5QR
w w w . c f z . o r g . u k**

Other books available from
CFZ PRESS

CFZ PRESS

ANIMALS & MEN - Issues 1 - 5 - In the Beginning
Edited by Jonathan Downes - ISBN 0-9512872-6-5

£12.50

At the beginning of the 21st Century monsters still roam the remote, and sometimes not so remote, corners of our planet. It is our job to search for them. The Centre for Fortean Zoology [CFZ] is the only professional, scientific and full-time organisation in the world dedicated to cryptozoology - the study of unknown animals. Since 1992 the CFZ has carried out an unparalleled programme of research and investigation all over the world. We have carried out expeditions to Sumatra (2003 and 2004), Mongolia (2005), Puerto Rico (1998 and 2004), Mexico (1998), Thailand (2000), Florida (1998), Nevada (1999 and 2003), Texas (2003 and 2004), and Illinois (2004). An introductory essay by Jonathan Downes, notes putting each issue into a historical perspective, and a history of the CFZ.

ANIMALS & MEN - Issues 6 - 10 - The Number of the Beast
Edited by Jonathan Downes - ISBN 978-1-905723-06-5

£12.50

At the beginning of the 21st Century monsters still roam the remote, and sometimes not so remote, corners of our planet. It is our job to search for them. The Centre for Fortean Zoology [CFZ] is the only professional, scientific and full-time organisation in the world dedicated to cryptozoology - the study of unknown animals. Since 1992 the CFZ has carried out an unparalleled programme of research and investigation all over the world. We have carried out expeditions to Sumatra (2003 and 2004), Mongolia (2005), Puerto Rico (1998 and 2004), Mexico (1998), Thailand (2000), Florida (1998), Nevada (1999 and 2003), Texas (2003 and 2004), and Illinois (2004). Preface by Mark North and an introductory essay by Jonathan Downes, notes putting each issue into a historical perspective, and a history of the CFZ.

BIG BIRD! Modern Sightings of Flying Monsters

Ken Gerhard - ISBN 978-1-905723-08-9

£7.99

From all over the dusty U.S./Mexican border come hair-raising stories of modern day encounters with winged monsters of immense size and terrifying appearance. Further field sightings of similar creatures are recorded from all around the globe. What lies behind these weird tales? Ken Gerhard is a native Texan, he lives in the homeland of the monster some call 'Big Bird'. Ken's scholarly work is the first of its kind. On the track of the monster, Ken uncovers cases of animal mutilations, attacks on humans and mounting evidence of a stunning zoological discovery ignored by mainstream science. Keep watching the skies!

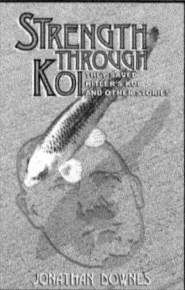

STRENGTH THROUGH KOI
They saved Hitler's Koi and other stories

£7.99

Jonathan Downes - ISBN 978-1-905723-04-1

Strength through Koi is a book of short stories - some of them true, some of them less so - by noted cryptozoologist and raconteur Jonathan Downes. The stories are all about koi carp, and their interaction with bigfoot, UFOs, and Nazis. Even the late George Harrison makes an appearance. Very funny in parts, this book is highly recommended for anyone with even a passing interest in aquaculture, but should be taken definitely *cum grano salis*.

CFZ PRESS, MYRTLE COTTAGE,
WOOLSERY, BIDEFORD,
NORTH DEVON, EX39 5QR

Other books available from
CFZ PRESS

BIG CATS IN BRITAIN YEARBOOK 2007
Edited by Mark Fraser - ISBN 978-1-905723-09-6

£12.50

People from all walks of life encounter mysterious felids on a daily basis, in every nook and cranny of the UK. Most are jet-black, some are white, some are brown; big cats of every description and colour are seen by some unsuspecting person while on his or her daily business. 'Big Cats in Britain' are the largest and most active research group in the British Isles and Ireland. This book contains a run-down of every known big cat sighting in the UK during 2006, together with essays by various luminaries of the British big cat research community.

CAT FLAPS! Northern Mystery Cats
Andy Roberts - ISBN 978-1-905723-11-9

£6.99

Of all Britain's mystery beasts, the alien big cats are the most renowned. In recent years the notoriety of these uncatchable, out-of-place predators have eclipsed even the Loch Ness Monster. They slink from the shadows to terrorise a community, and then, as often as not, vanish like ghosts. But now film, photographs, livestock kills, and paw prints show that we can no longer deny the existence of these once-legendary beasts. Here then is a case-study, a true lost classic of Fortean research by one of the country's most respected researchers.

CENTRE FOR FORTEAN ZOOLOGY 2007 YEARBOOK
Edited by Jonathan Downes and Richard Freeman
ISBN 978-1-905723-14-0

£12.50

The Centre For Fortean Zoology Yearbook is a collection of papers and essays too long and detailed for publication in the CFZ Journal *Animals & Men*. With contributions from both well-known researchers, and relative newcomers to the field, the Yearbook provides a forum where new theories can be expounded, and work on little-known cryptids discussed.

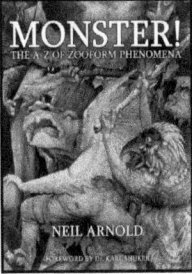

MONSTER! THE A-Z OF ZOOFORM PHENOMENA
Neil Arnold - ISBN 978-1-905723-10-2

£14.99

Zooform Phenomena are the most elusive, and least understood, mystery `animals`. Indeed, they are not animals at all, and are not even animate in the accepted terms of the word. Author and researcher Neil Arnold is to be commended for a groundbreaking piece of work, and has provided the world's first alphabetical listing of zooforms from around the world.

**CFZ PRESS, MYRTLE COTTAGE,
WOOLFARDISWORTHY BIDEFORD,
NORTH DEVON, EX39 5QR
www.cfz.org.uk**

Other books available from
CFZ PRESS

BIG CATS LOOSE IN BRITAIN
Marcus Matthews - ISBN 978-1-905723-12-6

£14.99

Big Cats: Loose in Britain, looks at the body of anecdotal evidence for such creatures: sightings, livestock kills, paw-prints and photographs, and seeks to determine underlying commonalities and threads of evidence. These two strands are repeatedly woven together into a highly readable, yet scientifically compelling, overview of the big cat phenomenon in Britain.

DARK DORSET
TALES OF MYSTERY, WONDER AND TERROR
Robert. J. Newland and Mark. J. North
ISBN 978-1-905723-15-6

£12.50

This extensively illustrated compendium has over 400 tales and references, making this book by far one of the best in its field. Dark Dorset has been thoroughly researched, and includes many new entries and up to date information never before published. The title of the book speaks for itself, and is indeed not for the faint hearted or those easily shocked.

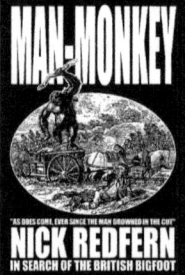

MAN-MONKEY - IN SEARCH OF THE BRITISH BIGFOOT
Nick Redfern - ISBN 978-1-905723-16-4

£9.99

In her 1883 book, *Shropshire Folklore*, Charlotte S. Burne wrote: *'Just before he reached the canal bridge, a strange black creature with great white eyes sprang out of the plantation by the roadside and alighted on his horse's back'*. The creature duly became known as the `Man-Monkey`.

Between 1986 and early 2001, Nick Redfern delved deeply into the mystery of the strange creature of that dark stretch of canal. Now, published for the very first time, are Nick's original interview notes, his files and discoveries; as well as his theories pertaining to what lies at the heart of this diabolical legend.

EXTRAORDINARY ANIMALS REVISITED
Dr Karl Shuker - ISBN 978-1905723171

£14.99

This delightful book is the long-awaited, greatly-expanded new edition of one of Dr Karl Shuker's much-loved early volumes, *Extraordinary Animals Worldwide*. It is a fascinating celebration of what used to be called romantic natural history, examining a dazzling diversity of animal anomalies, creatures of cryptozoology, and all manner of other thought-provoking zoological revelations and continuing controversies down through the ages of wildlife discovery.

**CFZ PRESS, MYRTLE COTTAGE,
WOOLFARDISWORTHY BIDEFORD,
NORTH DEVON, EX39 5QR
w w w . c f z . o r g . u k**

Other books available from
CFZ PRESS

DARK DORSET CALENDAR CUSTOMS
Robert J Newland - ISBN 978-1-905723-18-8

£12.50

Much of the intrinsic charm of Dorset folklore is owed to the importance of folk customs. Today only a small amount of these curious and occasionally eccentric customs have survived, while those that still continue have, for many of us, lost their original significance. Why do we eat pancakes on Shrove Tuesday? Why do children dance around the maypole on May Day? Why do we carve pumpkin lanterns at Hallowe'en? All the answers are here! Robert has made an in-depth study of the Dorset country calendar identifying the major feast-days, holidays and celebrations when traditionally such folk customs are practiced.

CENTRE FOR FORTEAN ZOOLOGY 2004 YEARBOOK
Edited by Jonathan Downes and Richard Freeman
ISBN 978-1-905723-14-0

£12.50

The Centre For Fortean Zoology Yearbook is a collection of papers and essays too long and detailed for publication in the CFZ Journal *Animals & Men*. With contributions from both well-known researchers, and relative newcomers to the field, the Yearbook provides a forum where new theories can be expounded, and work on little-known cryptids discussed.

CENTRE FOR FORTEAN ZOOLOGY 2008 YEARBOOK
Edited by Jonathan Downes and Corinna Downes
ISBN 978 -1-905723-19-5

£12.50

The Centre For Fortean Zoology Yearbook is a collection of papers and essays too long and detailed for publication in the CFZ Journal *Animals & Men*. With contributions from both well-known researchers, and relative newcomers to the field, the Yearbook provides a forum where new theories can be expounded, and work on little-known cryptids discussed.

ETHNA'S JOURNAL
Corinna Newton Downes
ISBN 978 -1-905723-21-8

£9.99

Ethna's Journal tells the story of a few months in an alternate Dark Ages, seen through the eyes of Ethna, daughter of Lord Edric. She is an unsophisticated girl from the fortress town of Cragnuth, somewhere in the north of England, who reluctantly gets embroiled in a web of treachery, sorcery and bloody war...

**CFZ PRESS, MYRTLE COTTAGE,
WOOLFARDISWORTHY BIDEFORD,
NORTH DEVON, EX39 5QR
www.cfz.org.uk**

Other books available from
CFZ PRESS

ANIMALS & MEN - Issues 11 - 15 - The Call of the Wild
Jonathan Downes (Ed) - ISBN 978-1-905723-07-2

£12.50

Since 1994 we have been publishing the world's only dedicated cryptozoology magazine, *Animals & Men*. This volume contains fascimile reprints of issues 11 to 15 and includes articles covering out of place walruses, feathered dinosaurs, possible North American ground sloth survival, the theory of initial bipedalism, mystery whales, mitten crabs in Britain, Barbary lions, out of place animals in Germany, mystery pangolins, the barking beast of Bath, Yorkshire ABCs, Molly the singing oyster, singing mice, the dragons of Yorkshire, singing mice, the bigfoot murders, waspman, British beavers, the migo, Nessie, the weird warbling whatsit of the westcountry, the quagga project and much more...

IN THE WAKE OF BERNARD HEUVELMANS
Michael A Woodley - ISBN 978-1-905723-20-1

£9.99

Everyone is familiar with the nautical maps from the middle ages that were liberally festooned with images of exotic and monstrous animals, but the truth of the matter is that the *idea* of the sea monster is probably as old as humankind itself.

For two hundred years, scientists have been producing speculative classifications of sea serpents, attempting to place them within a zoological framework. This book looks at these successive classification models, and using a new formula produces a sea serpent classification for the 21st Century.

CENTRE FOR FORTEAN ZOOLOGY 1999 YEARBOOK
Edited by Jonathan Downes and Corinna Downes
ISBN 978 -1-905723-24-9

£12.50

The Centre For Fortean Zoology Yearbook is a collection of papers and essays too long and detailed for publication in the CFZ Journal *Animals & Men*. With contributions from both well-known researchers, and relative newcomers to the field, the Yearbook provides a forum where new theories can be expounded, and work on little-known cryptids discussed.

CENTRE FOR FORTEAN ZOOLOGY 1996 YEARBOOK
Edited by Jonathan Downes and Corinna Downes
ISBN 978 -1-905723-22-5

£12.50

The Centre For Fortean Zoology Yearbook is a collection of papers and essays too long and detailed for publication in the CFZ Journal *Animals & Men*. With contributions from both well-known researchers, and relative newcomers to the field, the Yearbook provides a forum where new theories can be expounded, and work on little-known cryptids discussed.

**CFZ PRESS, MYRTLE COTTAGE,
WOOLFARDISWORTHY BIDEFORD,
NORTH DEVON, EX39 5QR
w w w . c f z . o r g . u k**

Other books available from
CFZ PRESS

BIG CATS IN BRITAIN YEARBOOK 2008
Edited by Mark Fraser - ISBN 978-1-905723-23-2

£12.50

People from all walks of life encounter mysterious felids on a daily basis, in every nook and cranny of the UK. Most are jet-black, some are white, some are brown; big cats of every description and colour are seen by some unsuspecting person while on his or her daily business. 'Big Cats in Britain' are the largest and most active research group in the British Isles and Ireland. This book contains a run-down of every known big cat sighting in the UK during 2006, together with essays by various luminaries of the British big cat research community.

**CFZ PRESS, MYRTLE COTTAGE,
WOOLFARDISWORTHY BIDEFORD,
NORTH DEVON, EX39 5QR
w w w . c f z . o r g . u k**